CIRCUIT ANALYSIS
Theory and Practice, 4E
&
CIRCUIT ANALYSIS with DEVICES
Theory and Practice, 2E

LABORATORY MANUAL

Allan H. Robbins
Wilhelm C. Miller

THOMSON

DELMAR LEARNING

Australia   Canada   Mexico   Singapore   Spain   United Kingdom   United States

THOMSON

DELMAR LEARNING

Lab Manual to Accompany Circuit Analysis: Theory and Practice, 4e and Circuit Analysis with Devices: Theory and Practice, 2e

Allan Robbins and Wilhelm Miller

Vice President, Technology and Trades ABU:

David Garza

Director of Learning Solutions:

Sandy Clark

Senior Acquisitions Editor:

Stephen Helba

Senior Product Manager:

Michelle Cannistraci

Marketing Director:

Deborah S. Yarnell

Channel Manager:

Dennis Williams

Marketing Coordinator:

Stacey Wiktorek

Editorial Assistant:

Dawn Daugherty

Senior Production Manager:

Larry Main

Production Editor:

Benj Gleeksman

Library of Congress Cataloging-in-Publication Data:
Card Number:

ISBN: 1418038644

NOTICE TO THE READER

Publisher does not warrant or guarantee any of the products described herein or perform any independent analysis in connection with any of the product information contained herein. Publisher does not assume, and expressly disclaims, any obligation to obtain and include information other than that provided to it by the manufacturer.

The reader is expressly warned to consider and adopt all safety precautions that might be indicated by the activities herein and to avoid all potential hazards. By following the instructions contained herein, the reader willingly assumes all risks in connection with such instructions.

The publisher makes no representation or warranties of any kind, including but not limited to, the warranties of fitness for particular purpose or merchantability, nor are any such representations implied with respect to the material set forth herein, and the publisher takes no responsibility with respect to such material. The publisher shall not be liable for any special, consequential, or exemplary damages resulting, in whole or part, from the readers' use of, or reliance upon, this material.

Contents

Preface

This manual provides a set of laboratory exercises that cover the basic concepts of circuit theory and electronic devices. While its 41 experiments are more than can be covered in a normal sequence in an introductory course, it provides flexibility and allows instructors to choose labs to suit their programs. The sequence is also flexible. For example, some instructors may wish to move the introductory oscilloscope lab to an earlier spot—immediately following Lab 8, for instance.

Each lab includes a short overview of key ideas, a set of objectives, a list of equipment and parts, instructions on how to carry out the investigation, appropriate tables for summarizing data, and a set of review questions or problems to test the student's comprehension of the material covered. A mini-tutorial and reference guide to equipment and basic measurement techniques is included at the beginning of the manual to provide the student with a readily accessible source of practical information for use during the lab program. A short section on safety in the laboratory follows this guide.

For users who require more emphasis on analog meters and measurements than was present in the 3rd edition of this manual, we have included a tutorial section entitled A Guide to Analog Meters and Measurements. It deals with analog meters, reading and interpreting VOM scales, meter errors, analog meter accuracy, and the like. Also reinstated (on the CD in the back of the text book) is a supplemental version of Lab 1 that includes laboratory practice to support this need.

The equipment needed to run these labs is, for the most part, common equipment of the type found at all colleges. Most experiments, for example, can be performed with just a variable dc power supply, digital multimeters, a signal generator, and a two-channel oscilloscope. A few require additional equipment such as an *LRC* meter (or an impedance bridge), a pair of wattmeters, and a three-phase source. A complete list of equipment and components needed is contained in Appendix B.

While this manual is designed as a companion lab book for the texts, *Circuit Analysis: Theory and Practice* and *Circuit Analysis with Devices: Theory and Practice*, by Allan H. Robbins and Wilhelm C. Miller, it may be used with any suitable text.

Allan H. Robbins
Wilhelm C. Miller
January 2006

Acknowledgments

We would like to thank a team at Thomson Delmar Learning for putting this project together. Specifically, we would like to mention Michelle Ruelos Cannistracci, Senior Project Manager, Benjamin Gleeksman, Production Editor, Larry Main, Senior Production Manager, Francis Hogan, Art and Design Coordinator, Dennis Williams, Marketing Manager, Dawn Daugherty, Editorial Assistant, Steve Helba, Senior Acquisitions Editor, and all their staffs. We also wish to thank the reviewers and users who have provided valuable feedback, as well as the people of Publishing Synthesis, New York, for their work in copyediting and preparing the manuscript for publication.

A Guide to Lab Equipment and Laboratory Measurements

Before we get started, let us take a brief look at lab equipment and basic measurement techniques. If you are unfamiliar with test equipment, you should read this section before you begin the lab program and reference it as necessary as you perform each lab.

The Power Supply

Your main source of dc in the laboratory will be a variable voltage, regulated dc power supply. The supply plugs into the ac wall outlet and converts the incoming ac to dc. Such supplies generally include a front panel control for setting the desired voltage and a built-in meter to monitor voltage (and sometimes current). A typical supply is shown in Figure 1. Both single-voltage and multi-voltage units are available. The typical range of output voltage for a laboratory supply is from zero to 30 volts.

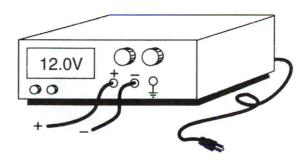

Figure 1   Power supply.

Floating Outputs

Most laboratory power supplies have *floating outputs*—that is, outputs where both the positive and negative terminals are isolated from ground. (The source voltage appears between the terminals, and

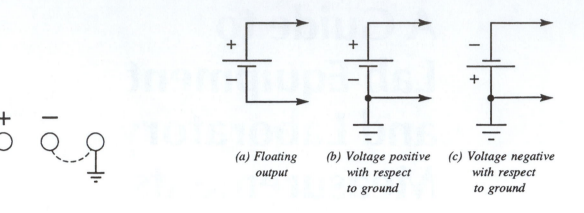

(a) Floating output

(b) Voltage positive with respect to ground

(c) Voltage negative with respect to ground

Figure 2   Either terminal may be jumpered to ground.

Figure 3   Possible connections

no voltage appears between either terminal and ground.) Such supplies may be operated with one or the other of its terminals jumpered to ground, Figure 2, or both terminals may be left floating. Possible connections are shown in Figure 3.

The Digital Multimeter (DMM)

Note
Until relatively recent years, analog multimeters were quite common. They have now been largely superseded by digital multimeters; however, there are situations where analog multimeters are useful. Accordingly, we have included a section entitled *A Guide to Analog Meters and Measurements* later in this guide (following the information on digital meters).

One of the most popular test instruments for basic measurements is the digital *multimeter* (DMM). It combines the functions of a voltmeter, ammeter, and ohmmeter into a single instrument that displays measured results on a numeric output. Figure 4 shows a hand-held DMM. The type of measurement to be made is selected by means of its function selector switch.

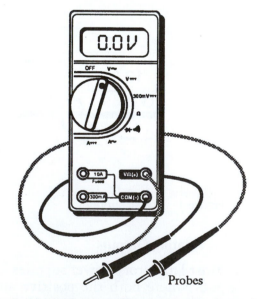

Figure 4   A handheld DMM. Reproduced with permission from Fluke Corporation.

Special Features of the DMM

Two features that most DMMs have are *autoranging* and *autopolarity*. With autoranging, you simply select the desired function (voltage, current, or resistance) then let the DMM automatically determine the correct range. Similarly for polarity—if you connect the instrument for a dc measurement, it automatically determines the polarity (for voltage) or direction (for current) and displays the appropriate sign as part of the measured result.

Manual Range Selection

For non-autoranging DMMs, you must manually select the appropriate range, e.g., 250 V full scale, 120 mA full scale, and so on. (Be sure that you do this before you energize the circuit.) If you have no idea of the magnitude of the quantity to be measured, start at the highest range to avoid possible instrument damage and work your way down to lower ranges until you get the best possible reading.

Terminal Connections

As indicated in Figure 4, a multimeter has two main terminals and several secondary terminals. One main terminal is generally designated *COMMON*, *COM* or minus (–). Designations for the other terminals vary. For example, some meters have a common set of terminals for voltage and resistance (which may be designated $V\Omega$ or plus (+) or some similar designation) plus one or more separate terminals for current. Other meters have a combined voltage/resistance/current input labelled $V\Omega A$. Standard lead colors are black and red. By convention, the black lead should be connected to the (–) or COM terminal while the red lead should be connected to the other main terminal, i.e., (+) or $V\Omega$.

Average Reading and True rms Reading Instruments

> **Note**
> For all labs in this manual, we will assume standard (i.e., non-true rms reading) DMMs with both autopolarity and autoranging.

AC meters are calibrated to read rms values. However, most meters (called *average responding*) are designed to measure rms for sine waves only i.e., they are not capable of determining the rms value for non-sinusoidal waveforms such as square waves, triangular waves, superimposed ac and dc, etc. For these, you need a special DMM called a *true rms* meter. True rms meters are somewhat more expensive than standard meters.

How To Measure dc Voltage

Refer to Figure 5: 1) Ensure that leads are plugged correctly into the meter sockets—red lead in the *V* socket and the black lead in the

COM socket; 2) Set the function selector to *dc voltage* and select the range if the meter is not autoranging; 3) Connect the probes across the circuit element whose voltage you wish to measure; 4) Read the voltage.

Points to Note with Respect to Figure 5: If the red (+) lead is connected to the positive side of the circuit and the black (–) lead to the negative side, the meter reading will be positive as in (a). Conversely, if the red lead is connected to the negative side of the circuit and the black lead to the positive side, the meter reading will be negative as in (b).

Reproduced with permission from Fluke Corporation

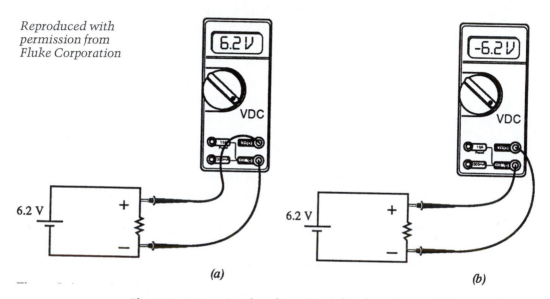

Figure 5   Measuring dc voltage. Note that the polarity of (b) is opposite that of (a)

How To Measure ac Voltage

The procedure for ac voltage is the same as for dc voltage except that you set the selector dial to *ac voltage*. In this case however, since an ac meter reads magnitude only, it does not matter which way you connect the meter. This is indicated in Figure 6. As you can see, both connections show the same result.

How To Measure dc Current

To measure dc current: 1) Turn power off, open the circuit, then connect so that the current you wish to measure passes through the meter as indicated in Figure 7. Ensure that the test lead is plugged into the current jack if the meter has a separate current input; 2) Set the function selector to dc current and select the range if necessary; 3) Energize the circuit and read the current.

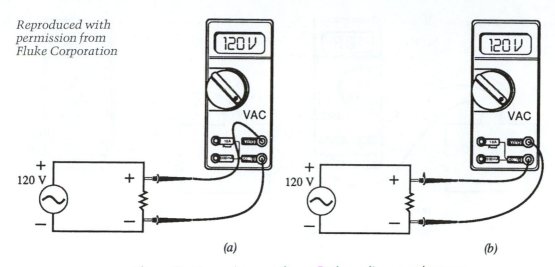

Reproduced with permission from Fluke Corporation

Figure 6   Measuring ac voltage. Both readings are the same.

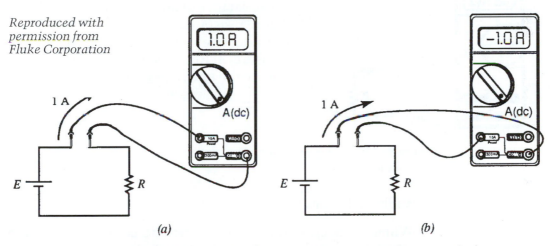

Reproduced with permission from Fluke Corporation

Figure 7   Measuring dc current. Be sure to use the current jack.

Points to Note: If the current enters at the A (or VΩA) terminal and exits at COM, the reading will be positive as in (a). If the leads are reversed as in (b), the reading will be negative.

How To Measure ac Current

Note: Not all multimeters measure ac current. For those that do, proceed as for the dc case, except set the dial to ac current instead of dc. (Since an ac meter reads magnitude only, you can connect as in Figure 8(a) or (b).

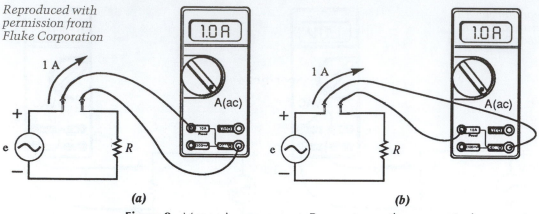

Reproduced with permission from Fluke Corporation

(a)   (b)

Figure 8   Measuring ac current. Be sure to use the current jack.

How To Measure Resistance

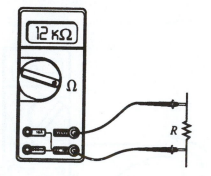

Figure 9 shows a DMM connected to measure the resistance of an isolated resistor. 1) Set the dial to resistance; 2) Connect the probe across the component whose resistance you wish to measure; 3) Read the measured value. Be sure to note the unit, Ω, $k\Omega$, etc.

Figure 9   Measuring resistance. Reproduced with permission from Fluke Corporation.

Additional Points to Note

1. When connecting the probes across the component to be measured, be careful not to touch both probe pins of the meter, since your body's resistance may introduce additional error. When measuring resistance, the red and black leads of the ohmmeter can usually be interchanged without affecting the reading. The exception is diodes and most active components such as transistors and integrated circuits (ICs).
2. When measuring the resistance of a resistor that is part of a circuit, disconnect the power from the circuit, else the ohmmeter may be damaged. In addition, it will be necessary to isolate the resistor from the rest of the circuit. You can do this by disconnecting at least one terminal of the component from the circuit.

Meter Errors

No meter can be guaranteed to be 100% accurate. Thus, when you measure a quantity, there will always be some uncertainty due to the meter itself. (This is similar to the situation with bathroom scales. If your scale is not perfectly accurate, it may indicate a few pounds higher or lower than your actual weight.)

Meter Accuracy Specifications

The *accuracy specification* for a meter defines the maximum error that it may have—that is, the guaranteed maximum value by which the measured value may deviate from the true value.

The basic dc voltage accuracy specifications for bench/portable and hand-held DMMs ranges from about 0.05% to 0.5%, depending on make and model. This specification is independent of reading— thus, a DMM with a 0.5% error specification will have no more than 0.5% error no matter what value it is reading. For example, if its displayed value is 100.0 volts, it could have up to 0.5% x 100 volts = 0.5 V error, meaning that the true value of the measured voltage can be anywhere between 99.5 and 100.5 volts. The same meter indicating 16.00 volts still has a possible 0.5% error of its reading (which in this case is $0.5\% \times 16$ volts = 0.08 V) and thus, its true value could be anywhere between 15.92 and 16.08 volts.

Some DMMs also have an additional uncertainty in their least significant digit. For example, a certain popular low cost, hand-held DMM has a specification of ±(0.5%+1 digit) which means that its maximum error at any point will not exceed one half a percent of its indicated reading plus one least significant digit. Thus if this meter is displaying 100.0 volts, it could have an error of half a volt (as noted above) plus 1 digit. This means that the true value of the measured voltage lies somewhere between 99.4 and 100.6 volts.

Current Measurement Accuracy Versus Voltage Measurement Accuracy

Depending on the meter you use, the accuracy of your current measurements may be poorer than the accuracy of your voltage measurements. For example, for the DMM mentioned above, the accuracy specification drops to ± (1.5% of reading + 2 digit) for current. On the other hand, some meters have the same accuracy specification for current as for voltage. Check your meter manual.

A Final Note on Meter Error Specifications

Accuracy specifications define a meter's worst case error and most meters will perform better than their published accuracy specifications. Still, the published figure is the only guarantee that you get. Also note that accuracy specifications for ac measurements may be different than for dc. Always check your meter manual.

Frequency Considerations

The frequency range of a DMM is an important consideration for ac measurements. For example, some DMMs can measure up to only about 1 kHz, while others can measure up to 100 kHz or higher. (If you need to make measurement at higher frequencies, you may have to use an oscilloscope, considered later.)

Schematic Symbols for Meters

So far, we have shown meters in pictorial form so that you can see how the meters are physically connected into a circuit and how the function selector switch is used. In practice, this is cumbersome and meters are usually represented by simple schematic symbols as in Figure 10. (In most labs, we show only the schematic. However, in the first few labs, we sometimes show both.)

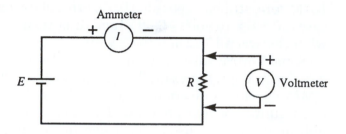

Figure 10   Representing meters using schematic symbols

Words of Caution

1. Never connect an ammeter across a voltage source. An ammeter is virtually a short circuit and damage to the meter or the source may occur.
2. For safety, always plug the probe leads into the meter sockets before connecting the probe tips to the circuit.
3. When using a meter with manual range selection, set the meter to its highest range if you do not know the approximate value of the quantity to be measured, then switch to lower ranges until you get to a suitable range.

The Function Generator

A *function generator*, Figure 11, is a variable frequency, multi-waveform source that produces a variety of waveforms such as sine, square, triangular, pulse, and so on. It will have a set of push-buttons or a selector switch for selecting the desired waveform, range selector push-buttons or a selector switch for setting the frequency range, a variable control for adjusting the frequency within the selected range, an amplitude control for adjusting the output voltage, and a digital readout (or an analog scale on the dial) for displaying the chosen frequency. A *dc offset* control is usually included so that you can add a positive or negative dc voltage component to the waveform. Depending on the make and model, other features may also be included. The usual frequency range of a function generator is from a few Hz to a few MHz.

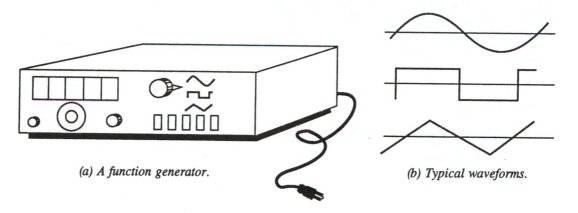

(a) A function generator.   (b) Typical waveforms.

Figure 11   A function generator and some of its waveforms

The Oscilloscope

The *oscilloscope* is an electronic test and measurement instrument that displays waveforms on a screen. It is the basic tool used for studying time-varying phenomena such as voltages and currents in electric circuits. It is used, for example, to measure the frequency and period of repetitive waveforms, to determine the rise and fall times of pulses, to find the phase difference between sinusoidal signals, to help troubleshoot electronic equipment, and so on. However, the oscilloscope is a rather complex instrument. We will therefore look at it in several stages. In this section, we look at it from a conceptual viewpoint; in later sections, we learn how to use it to make voltage, current, phase angle, and other measurements in a lab setting. Figure 12 depicts a typical dual-channel oscilloscope.

The Display Screen

The display screen of a standard oscilloscope is 10 cm (about four inches) across and is ruled with a grid at 1 cm intervals with graduated markings (called *graticules*) between grid lines. Measurements are made by positioning waveforms on this grid and reading values from its scale as illustrated in Figure 13. By selecting appropriate vertical and horizontal scale factors, you can determine the amplitudes of the waveforms, their period, and the phase displacement between them as we describe here and in later labs.

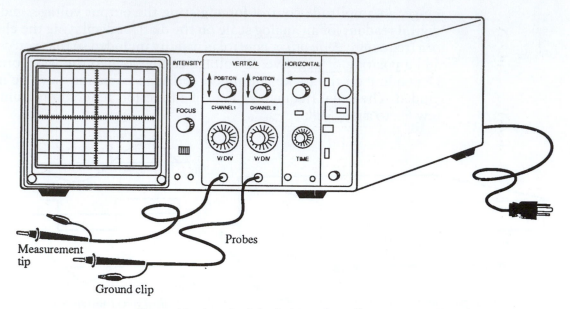

Figure 12   A typical dual-channel oscilloscope. Oscilloscopes are also referred to as scopes.

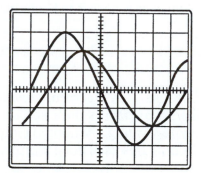

Figure 13   Waveforms on the screen.

Inside the Oscilloscope

Figure 14 is a simplified block diagram of an oscilloscope. It consists of a *CRT* (cathode ray tube), a *vertical display section*, a *horizontal display section* and *trigger circuits*.

The CRT

The heart of the oscilloscope is its CRT. (It displays the waveform.) The CRT is a vacuum tube, one end of which is flared out to make the screen. The screen is coated on the inside with a thin layer of phosphorescent material, and an *electron gun* inside the tube produces a stream of electrons and shoots them onto this screen. As each electron strikes the screen, it produces a spot of light. A horizontal deflection system then sweeps this spot across the face of the screen at

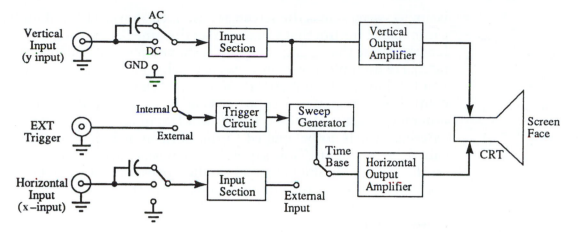

Figure 14 Simplified block diagram of an oscilloscope

a rapid rate, producing a horizontal line. This horizontal line can be positioned by means of a *vertical position control* located on the front panel. With no input signal present, you will usually position this line at the center of the screen.

The Vertical Deflection System

When an input signal is applied to the vertical input, the *vertical deflection system* causes the spot to move vertically in proportion to the voltage applied. With the spot moved vertically in response to the input signal and horizontally by the deflection system, a path is traced on the screen corresponding to the input waveform. If a dual-trace scope is used, two waveforms can be displayed simultaneously as in Figure 13. A vertical input is sometimes called a "y" input.

Because input signals may be large or small, an input *attenuator* and an *amplifier* are provided. A front panel control permits you to set the attenuation/amplification to suit the measurement. This control is calibrated in volts per division (VOLTS/DIV). A variable control allows for fine adjustment, but it is not calibrated. A switch on the input line allows you to select *ac coupled* or *dc coupled*. The switch also includes a GND (ground) position as indicated in Figure 14. When set to *dc*, the input signal is applied directly to the vertical deflection system, permitting the entire signal (ac and any dc component present) to be viewed on the screen; when set to *ac*, the input signal passes through a capacitor which blocks all dc and results in only the ac signal being displayed. The GND position permits you to establish a 0-V baseline by internally applying zero volts to the vertical deflection system.

The Horizontal Deflection System

The *horizontal deflection system* may be operated in one of two modes. In the *time-base* mode, an internal sweep circuit causes the beam to sweep across the screen as described above. This circuit has a front panel control that you use to set the time per division (i.e., the length of time that it takes the spot to move from one grid line to the next). For example, if you are measuring a high frequency waveform, you need to set the horizontal sweep at a high rate. Typically, this control is calibrated in seconds, ms and μs per screen division. The other mode of operation is the *horizontal input mode* which permits external signals to be applied to the horizontal input (sometimes called the "x" input). This is useful if you wish to display an x-y trace on the screen. On dual-channel oscilloscopes, one of the channels normally doubles as the x-input when used in the x-y mode.

Trigger Controls

To display a periodic waveform, the spot is swept repeatedly across the screen. The job of the *trigger circuit* is to always start the trace at the same point on the waveform so that each succeeding trace is superimposed on the preceding one, making it appear stationary on the screen. The trigger circuit synchronizes this process. As indicated in Figure 14, it samples the vertical input to determine when to start the trace moving horizontally across the screen. Two front panel controls, the *trigger level* and the *trigger slope* control permit you to select the trigger point. If triggering is not set properly, the trace will not be stable on the screen.

Sophisticated oscilloscopes have additional triggering controls. We will not consider them here.

Other Controls

Other controls include a *focus control* and an *intensity control*. These are used to adjust the trace for a sharp, crisp trace at a comfortable viewing level. A *horizontal position control* permits you to adjust the trace horizontally. A *beam finder* helps you locate the trace if it is off the screen and you are having trouble finding it.

Frequency Range

Oscilloscopes have an extremely broad frequency range; even low-cost models can measure signals from dc to 20 MHz. Top of the line models can measure signals up to several hundred MHz and digital oscilloscopes can measure up into the GHz range.

Probes

Probes are an integral part of a scope's measurement system. Probes are either "× 1" or "× 10" (times one or times ten). A "× 1" probe has no attenuation and thus, the full voltage at its tip is applied to the scope. A "× 10" probe on the other hand, has a built-in ten-to-one voltage divider, and thus, only one tenth of the measured voltage is applied to the scope. With a "× 1" probe, the input resistance of the scope is 1 megohm; with a "× 10" probe, it is 10 megohms. Thus, the "× 10" probe results in less loading of high impedance circuits.

Each probe has a built-in *compensation circuit* which must be "tuned" to match the channel to which it is connected. Depending on the manufacturer, this is done by twisting the barrel of the probe or by adjusting a compensator screw on the probe. To tune the compensation circuit, touch the probe to the calibration test point on the front panel of the scope and adjust until the waveform seen on the screen is a flat-topped square wave. Once the probe is tuned, it will not introduce distortions into the measurements that you are making.

Final Comments

Although all oscilloscopes have the above basic features in common, there are differences in detail that you will have to learn from your instructor and from the scope's manual. Much of this you will learn as you perform the labs.

A Guide to Analog Meters and Measurements

Although digital multimeters are more widely used than analog multimeters, analog multimeters are still found in use. And in fact, there are situations where analog meters are preferable to digital ones. One of these is in observing trends—i.e., slowly varying voltages or currents. It is easier to observe and make sense of a slowly swinging needle indication than it is to glean the same information from a set of digits rolling by on a digital readout. In this guide, we look at analog meters, their use, their errors, and how to interpret their scale readings.

VOMs

Analog multimeters are frequently referred to as VOMs, since they are multipurpose instruments that can be used to measure volts, ohms or milliamps. Figure 15 shows a typical meter. As can be seen, it uses a needle pointer and a set of scales to display measured values. A function selector switch provides the means to select the quantity to be measured and to specify the scale to be read. Note that the range selector switch must be manually operated; VOMs do not have autoranging capabilities as DMMs have. You must set the selector switch to the appropriate range—e.g., 250 V full scale, 120 mA full scale—before you connect the meter. (This is discussed in more detail later.)

Terminal Connections

Like a DMM, a VOM has two main terminals and one or more secondary terminals. One main terminal is generally designated *COMMON, COM* or minuts (–). Designations for the other terminals vary. For example, some meters have a common set of terminals for voltage and resistance (which may be designated VΩ or plus (+) or some similar designation) plus one or more separate terminals for current. Other meters have a combined voltage/resistance/current input labelled *VΩA*. Standard lead colors are black and red; the black lead connects to the (–) or COM terminal while the red lead connects to the other main terminal, i.e., (+) or *VΩ*.

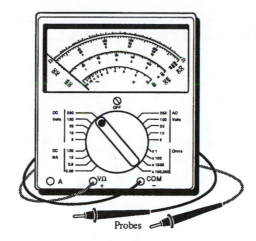

Figure 15   An analog VOM

How to Measure Voltage, Current, and Resistance with a VOM

The same basic principles apply to measuring voltage, current and resistance when using a VOM as when using a digital DMM, so consult the diagrams and discussions in the DMM sections earlier. However, here are some important differences to note:

- Analog VOMs are not auto-ranging: you must select the range yourself. If you don't know the approximate value of the quantity to be measured, set the meter to the highest range possible (before connecting the meter), then switch to successively lower ranges until you get a good reading
- Analog VOMs do not have autopolarity. The pointer rotates upscale, but if you connect a meter backward (for dc measurements), the pointer attempts to pivot downscale, but is unable to make a measurement because it comes up against the end stop. In this case, you must reverse (that is, correct) the probe connections at the point of measurement
- For ac measurements, polarity is immaterial and the previous point does not apply
- VOMs have lower impedance than DMMs; hence, they load circuits more than do digital meters
- Compared to typical digital DMMs, analog multimeters have higher frequency responses, typically up to about 100 kHz
- While analog VOMs are calibrated to read rms for ac sine waves, they are generally average-responding meters. This means that they cannot measure the rms value for any other waveform

Interpreting Analog Meter Scales

Analog instruments have multiple scales, and values are somewhat more difficult to interpret than for digital instruments. To interpret

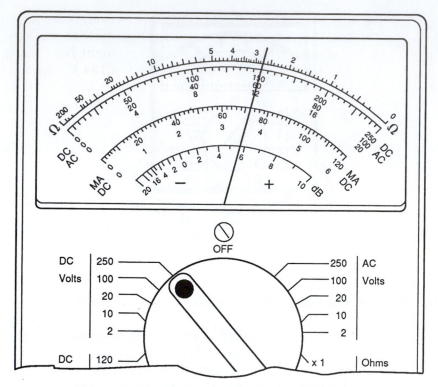

Figure 16   Reading an analog sale. V = 150 V.

a result, you must note both the scale reading and the range to which the function selector switch is set. For example, in Figure 16, the dial is set to the 250-V range and you must therefore read the voltage on the 250-V scale. Here, V = 150 V.

Note that analog scales are divided into intervals (called *primary divisions*) with unmarked graduations (called *secondary divisions*) between. For example, on the 250-V range of Figure 16, each primary division represents 50 V. By counting the number of secondary divisions between 50-V markings, you can see that each represents an increment of 5 volts. Reading a value is straightforward if the needle is over a primary division as indicated in Figure 16. Reading a voltage is also straightforward if the needle rests over a secondary division as in Figure 17(a). By counting the number of 5-V increments, you can

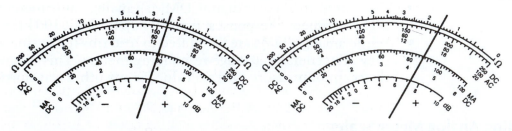

(a) V = 160 V.   *(b) There is some uncertainty here.*

Figure 17   Interpreting meter scales. The meter here is set to the 150-V range.

see that the meter indicates 160 V. However, if the needle rests part way between graduations, you must estimate the reading. For example, the reading shown in Figure 17(b) might be interpreted by one person as 183 V but by another person as 184 V. The estimated digit will have some uncertainty due to your inability to read it exactly.

Range Selection

As noted, some scales are used by more than one range. For example, in Figure 16 the 100-V scale is used by both the 100-V and the 10-V ranges. When the selector switch is set to 100 V, full-scale voltage is 100 volts and the scale is read directly, with each secondary division representing 2 volts. However, when it is set to 10 V, full-scale voltage is 10 volts and readings must be divided by 10. In this case, secondary divisions represent 0.2 volts. Thus, for Figure 16, if the range selector were set to 100 V, the reading would be 60 V, while if it were set to 10 V, the reading would be 6 volts.

Meter Errors

No meter can be guaranteed to be 100% accurate. Thus, when you measure a quantity, there will always be some uncertainty due to the meter itself. (This is similar to the situation with bathroom scales. If your scale is not perfectly accurate, it may indicate a few pounds higher or lower than your actual weight.) In general, DMMs are more accurate than analog VOMs.

Meter Accuracy Specifications

The *accuracy specification* for a meter defines the maximum error that it may have—that is, the guaranteed maximum value by which the measured value may deviate from the true value.

The accuracy of a VOM is defined in terms of its full scale value. Thus a VOM with an accuracy specification of 2% measuring a voltage on a 50-V scale may have as much as 1 volt error anywhere on that scale (since 2% of 50 V is 1 V). While this 1-volt error represents 2% at full scale, it represents 4% at half scale and 10% at the 10-V point. Thus, for a measured value of 50 volts, the true value may lie anywhere between 49 and 51 volts, but for a measured value of 10 volts, the true value may be anywhere between 9 and 11 volts. Accuracy can be improved by changing to a range where the reading is closer to full scale. For example, a 2% error on the 10-V range represents a possible error of 0.2 volts; thus, if you measure 10 volts on the 10-V range, the true value lies between 9.8 and 10.2 volt, a considerable improvement over the same reading taken on the 50-V scale. (It is for this reason that you should select a range that yields a reading close to full scale when using a VOM.) In addition, VOM accuracies are usually specified with the meter lying in a horizontal position; if the meter is vertical, its accuracy could be poorer due to the construction of the needle pivot. It may help to tap the meter lightly if the needle appears stuck.

A Final Note

Accuracy specifications for ac measurements are generally poorer than for dc measurements. (Check the operator's manual for your particular meter for details.) Note also that accuracy specifications define a meter's worst-case error and that most meters will perform better than their published accuracy specifications for many measurements. Still, the published figure is the only guarantee that you get. Note also that manufacturers may guarantee their specifications for a limited time only (e.g., one year) after calibration.

Safety

Everyone has handled electrical appliances and tools in one form or another and most have never experienced a serious injury or mishap involving electricity. As a result, most of us think of electricity as a safe form of energy; if we think of it at all! However, we have all heard of someone getting a nasty shock as the result of carelessness or from poorly maintained equipment.

Although it is impossible to foresee every possible injury, most electrical accidents are preventable by using common sense and by adopting a heathy respect for electrical energy.

Current kills!

The human body uses electrical impulses as low as 10 mV (ten one-thousandths of a volt) to transmit nerve messages between the brain and the various parts of the body. When an external source of electricity interacts with the body, the results can be disastrous! The human body consists mostly of water, with numerous dissolved electrolytes such as potassium chloride, phosphates and sodium chloride, making us very good conductors of electricity. An alternating current as low as 5 mA (five one-thousandths of an ampere) can be quite painful, causing muscles to go into spasms. When the current is increased to 10 mA, the spasms may be sufficient to prevent the victim from letting go of the current source. If the current is increased to 15 mA, breathing may be stopped and the heart itself may go into spasms (fibrillation) preventing the flow of blood to the brain. If the victim does not receive immediate attention, death may result.

In the event of an electrical accident in the lab, there are some basic steps that you should take.

1. **Turn the power off if you can safely do so, otherwise (if you can safely do so), remove the victim from contact with the source of electricity**. Since touching the victim also puts the rescuer at risk, it is necessary to do this without coming into electrical contact. One way to do this is to use an insulated object such as a dry broom or other long, dry stick.
2. **Call for professional help**.
3. **If you are qualified, you can administer first aid**. You will need to tell the attending paramedics how the victim arrived at his/her injuries and what first aid was administered.

Preventing Electrical Injury

Rather than treating an electrical injury, it best to prevent the occurance in the first place. The following is a list of safety rules which should be observed in the workplace or the lab.

1. **Consider all voltages above 50 Vdc or 30 Vac (rms) as a potential hazard**. If a voltage source is touched with dry hands you may experience only a mild tingle. The same source may be painful or fatal if your skin is moist or has a cut. If you experience even a minor electrical shock, mention this to your instructor or lab supervisor so that no one else is subjected to a potential hazard. It may be necessary to service the equipment.

2. **Ensure that all equipment is operating correctly and that no power cords are frayed or cut**. If a power cord is not in perfect condition, label the fault and bring it to the attention of your instructor or supervisor. Do not put a faulty piece of equipment into service. Tag the equipment and submit it for repair.

3. **Carefully insert and remove power cords**. Remove power cords by pulling on the connector. Never remove a power cord from the socket by pulling on the cable. There is a good chance that the cable will eventually tear free of the connector, presenting a fire or electrical hazard.

4. **Consider all sources of RF** (radio frequency) **energy as potential hazards**. Signals at very high frequencies (such as microwaves) have the potential to damage molecular structures (including your body's skin and internal organs) without immediately apparent symptoms.

5. **Use electrolytic capacitors with caution**. Some electrolytic capacitors are able to store large amounts of charge and may have the capacity to inflict the same injuries as voltage sources. Ensure that electrolytic capacitors are connected into the circuit with the correct polarity and verify that the voltage rating of the capacitor is not exceeded. If an electrolytic capacitor is connected incorrectly, the result may be an explosion, sending electrolytic chemicals and sharp pieces of metal into the air. Safety glasses provide some protection from shrapnel and chemical spills.

6. **Use the correct protective equipment**. Prior to working on circuits with dangerous voltages, interrupt the circuit at the control panel by opening the switch. To prevent someone from inadvertently closing the switch, tag the circuit at the control panel, clearly stating the reason for the interrupt. Wear the appropriate protective gear such as safety glasses, gloves and boots. Avoid wearing jewelry and loose fitting clothing. Use a rubber mat and rubber-soled boots to work on circuits in a damp environment.

7. **Know the location of fire extinguishers, circuit breakers, and panic safety switches**. In the event of an accident, the power should be immediately turned off. If a fire has broken out, you

may need to use a fire extinguisher. Ensure that the extinguisher is appropriate for the particular application. A fire should only be fought if it is not between you and a safe exit and if there is a good chance that you can bring the fire under control.

8. **Follow your instructor's safety instructions**. Since each lab has its own procedures which need to be followed, your instructor (or lab supervisor) will provide you with special precautions for the equipment or components used.

A Note Concerning Calculated and Measured Results

In the real world, calculated results and measured results seldom agree exactly. It is important to understand that there are usually explainable reasons for the differences, such as component tolerances, inherent meter errors, human measurement errors and so on. As you analyze the data for each experiment, you should try to determine the explainable differences, and discuss them as part of the analysis for the lab.

Trademarks

MultiSIM is a registered trademark of Electronics Workbench, a National Instruments company.

OrCAD PSpice is a registered trademark of Cadence Design Systems, Inc.

Powerstat® is a registered trademark of *The Superior Electric Co.*

Summary of Some Safety Precautions When Using Lab Equipment

Here are some general safety precautions to be observed for *all* lab work. You will also find specific safety advice in later labs. At all times, be careful.

1. Turn power off before making changes to your circuit.
2. Insert the probes into the meter before connecting the probes to your circuit in order to avoid "live" dangling leads.
3. Select the function and range (if necessary) of your meters before energizing your circuit. If you do not have autoranging and don't know the approximate voltage or current that you are about to measure, set the meter to its highest range and work your way down to the appropriate range.
4. Keep your fingers behind the finger guards on the probes when making measurements. (Although this is not essential for low voltage circuits, it is wise to develop safe work habits from the outset.)

Voltage and Current Measurement and Ohm's Law

OBJECTIVES

After completing this lab, you will be able to
* measure voltage, current and resistance in a dc circuit,
* confirm Ohm's law by direct measurement.

EQUIPMENT REQUIRED

☐ Two DMMs
☐ Variable dc power supply

COMPONENTS

☐ Resistors: One 1-kΩ resistor, 1/4-W, 5% tolerance or better
Two 2-kΩ resistors, 1/4-W, 5% tolerance or better

PRELIMINARY

Before you start, review the section on measuring voltage, current and resistance in *A Guide to Lab Equipment and Laboratory Measurements* at the front of this manual.

EQUIPMENT USED

Instrument	Manufacturer/Model No.	Serial No.
DMM #1		
DMM #2		
Power supply		

Table 1-1

TEXT REFERENCE

Section 4.1   OHM'S LAW
Section 4.7   Nonlinear and Dynamic Resistance

DISCUSSION

The most fundamental relationship of circuit theory is Ohm's law. Ohm's law states that in a purely resistive circuit, current is directly proportional to voltage and inversely proportional to resistance. In equation form

$$I = \frac{V}{R} \text{ (Ohm's law)} \tag{1-1}$$

where V is in volts, R is in ohms, I is in amps and reference conventions for voltage and current are as shown as in Figure 1-1. This relationship (which holds for every resistance in a circuit) is an experimental result and we will thus investigate it experimentally. First, you need to learn how to measure voltage and current.

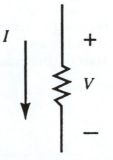

Figure 1-1   Conventions for Ohm's law.

DC Voltage Measurement

The basic voltage measuring scheme is shown in Figure 1-2. Be sure to adhere to the lead-color convention depicted and the safety checklist shown.

Current Measurement

The basic test circuit is depicted in Figure 1-3. When connected as shown, the DMM will yield a positive reading, whereas, if the test leads are reversed (so that current enters the COM terminal), the DMM will yield a negative reading.

MEASUREMENTS

PART A: Fixed Resistance, Variable Voltage

1. Measure the value of each resistor and record in Table 1-2. (Mark each 2-kΩ resistor with masking tape so that you can keep track of them.)

2. With power off, select R_2 and assemble the circuit as in Figure 1-4. Be sure to connect the meters so that they indicate positive values.

	Nominal	Measured
R_1	1-kΩ	
R_2	2-kΩ	
R_3	2-kΩ	

Table 1-2

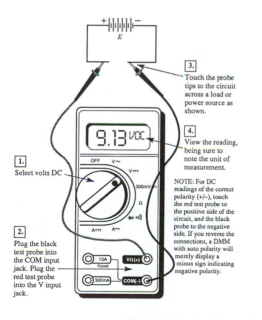

1. Select volts DC

2. Plug the black test probe into the COM input jack. Plug the red test probe into the V input jack.

3. Touch the probe tips to the circuit across a load or power source as shown.

4. View the reading, being sure to note the unit of measurement.

NOTE: For DC readings of the correct polarity (+/−), touch the red test probe to the positive side of the circuit, and the black probe to the negative side. If you reverse the connections, a DMM with auto polarity will merely display a minus sign indicating negative polarity.

Safety Checklist
- Always turn power off before making changes to your circuit.
- Select the function (and range if necessary) before connecting the meter. If you do not have autoranging and you don't know the approximate voltage, begin with the highest range and work your way down.
- Keep your fingers behind the finger guards on the probes when making measurements. (Although this is not essential here because of the low voltages used, it is a good idea to develop safe work habits from the outset.)

Figure 1-2 Making a dc voltage measurement with a DMM. Steps 1 and 2 may be reversed. (Adapted from the ABCs of DMMs. Courtesy of Fluke Corporation. Used with permission.)

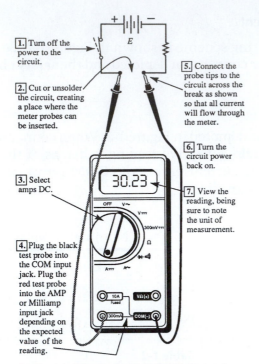

1. Turn off the power to the circuit.

2. Cut or unsolder the circuit, creating a place where the meter probes can be inserted.

3. Select amps DC.

4. Plug the black test probe into the COM input jack. Plug the red test probe into the AMP or Milliamp input jack depending on the expected value of the reading.

5. Connect the probe tips to the circuit across the break as shown so that all current will flow through the meter.

6. Turn the circuit power back on.

7. View the reading, being sure to note the unit of measurement.

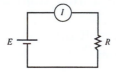

(a) Pictorial. Although a hand-held DMM is illustrated a bench-top DMM or a VOM may be used instead.

(b) Schematic

> **Meter Terminal Designations**
> While the DMM shown has separate terminals for current and voltage, not all meters follow this practice. Some meters for example use a combined input designated VΩA, which means that the current input terminal is the same as the voltage input terminal. In this case, when you set the selector switch to A (or amps), the meter functions as an ammeter, whereas when you set it to V, it functions as a voltmeter. As always, connect the black lead to the COM terminal and the red lead to the other terminal.

Figure 1-3   DC current measurement. (Part (a) adapted from The ABCs of DMMs. Courtesy of Fluke Corporation. Used with permission.)

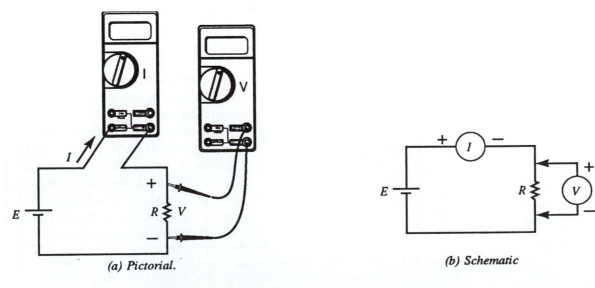

(a) Pictorial.

(b) Schematic

Figure 1-4   Connections for studying Ohm's Law.

a. Set the voltage to $V = 8$ V and measure the current.

   $I =$ _____ (Record this result also in Table 1-3.)

b. Set the voltage to $V = 16$ V and measure the current.

   $I =$ _____

c. Set the voltage to $V = 4$ V and measure the current.

   $I =$ _____

d. Based on these observations, for a fixed resistance, how does current vary with voltage (within the limitations of accuracy of your meters and components)?

_____

PART B: Fixed Voltage, Variable Resistance

3. a. Turn off the power supply. Using both 2-kΩ resistors, connect the circuit as in Figure 1-5. (This doubles the circuit resistance to 4 kΩ.) Set $V = 8$ volts and measure the current.

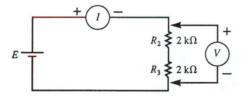

Figure 1-5   Doubling the circuit resistance. Here, $R_T = 4$ kΩ.

   $I =$ _____ . (Record also in Table 1-3.)

b. Turn off the power supply. Using the 1 kΩ resistor, reconnect the circuit as in Figure 1-4. Set $V = 8$ volts and measure the cur-

   rent. $I =$ _____ . (Record also in Table 1-3.)

c. Consider Table 1-3. Within the accuracy of the results obtained, for a fixed voltage, how does current vary with resistance?

_____

	V = 8 V	
Test	R	I (mA)
2(a)	2 kΩ	
3(a)	4 kΩ	
3(b)	1 kΩ	

Table 1-3

4. For Tests 2 and 3, compute current using Ohm's law and tabulate in Table 1-4. (Use the measured values of resistance for each case.) If there are differences between calculated and measured current, what is the likely cause? _____

_____

			Current I (mA)	
Test	V	R	Calculated	Measured
2(a)	8 V			
2(b)	16 V			
2(c)	4 V			
3(a)	8 V			
3(b)	8 V			

Table 1-4

PART C: Ohm's Law Graph

5. Using the circuit of Figure 1-4 and the 1-kΩ resistor, vary V from 0 volt to 10 volts in 2 volt increments. Tabulate your results in Table 1-5 and plot the results on the graph of Figure 1-6. Replace R with a 2-kΩ resistor and repeat.
6. Resistance can be calculated at any point on the Ohm's law graph. For the nominal 1-kΩ resistor, at V = 5 volts, determine current from Figure 1-6 and compute R using Equation 1-1. How well does it agree with the value of R used in Test 5? Repeat for the nominal 2-kΩ resistor.

_____

_____

_____

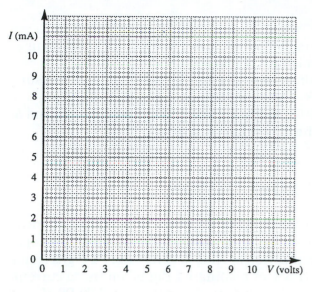

	Current I (mA)	
V	1-kΩ	2-kΩ
0 V		
2 V		
4 V		
6 V		
8 V		
10 V		

Figure 1-6 Ohm's law Graph.

7. Resistance may be determined from a *V-I* plot as described in Sec. 4.7 of the text. Select an arbitrary value for ΔV, then from the graph for the 1-kΩ resistor of Fig. 1-6 determine the resulting ΔI. Using these values, compute resistance. Show all work, including a sketch with ΔV and ΔI clearly marked.

COMPUTER ANALYSIS

8. Using either MultiSIM or PSpice as applicable set up the circuit of Fig. 1-4. Repeat PART A and PART B of the lab and compare to your measured results.

PROBLEMS

9. A 12-V source is applied to a resistor with color code orange, black, brown. What is the current in mA?
10. Make a sketch to show how to connect voltmeters to measure the voltage across R_2 and R_3 of Fig. 1-5.
11. For Figure 1-4, $I = 10$ mA. If E is increased by a factor of 16 and R is halved, what is the new value of current?

FOR FURTHER INVESTIGATION AND DISCUSSION

Write a short discussion paper on the impact of meter errors on measured values. As the basis for your discussion, consider a 4-digit DMM with an error specification of ±(0.5% of reading + 1 digit) on its dc voltage ranges and ±(1.5% of reading + 2 digits) on its dc current ranges. Consider Figure 1-4. Suppose the meters read 24.00 V and 12.00 mA respectively. Compute the nominal value for resistance (assuming no meter errors) as well as the minimum and maximum resistance values assuming maximum meter errors. Based on this, what can you conclude about accepting results (derived from measurements) at face value?

LAB 2

Series dc Circuits

OBJECTIVES

After completing this lab, you will be able to
- assemble a series circuit consisting of a voltage source and several resistors,
- use a DMM to measure voltage and current in a series circuit,
- compare measured values to theoretical calculations and verify Kirchhoff's voltage law,
- measure the effects of connecting several voltage sources in series,
- connect a circuit to ground using the ground terminal of a voltage source and use a DMM to measure voltages between several points and ground,
- measure the internal resistance of several voltage sources.

EQUIPMENT REQUIRED

☐ Digital multimeter (DMM)
☐ dc power supply (2)
  Note: Record this equipment in Table 2-1.

COMPONENTS

☐ Resistors: 47-Ω, 100-Ω, 270-Ω, 330-Ω, 470-Ω (1/4-W)
  47-Ω, 100-Ω, 270-Ω (2-W)
☐ Batteries: 1.5-V D-cell, 9-V (MN 1604 or equivalent)

EQUIPMENT USED

Instrument	Manufacturer/Model No.	Serial No.
DMM		
dc Supply		
dc Supply		

Table 2-1

TEXT REFERENCE

DISCUSSION

Two elements are said to be in *series* if they are connected at a single point and if there are no other current-carrying connections at this point. Each element in a series circuit has the same current as, illustrated in Figure 2-1.

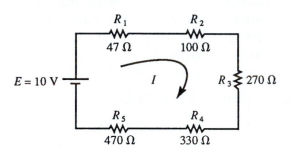

Figure 2-1 Series circuit

The equivalent resistance of n resistors in series, is determined as the summation

$$R_T = R_1 + R_2 + ... + R_n \qquad (2\text{-}1)$$

When these resistors are connected in series with a voltage source, the current in the circuit is given as

$$I = \frac{E}{R_T} \qquad (2\text{-}2)$$

The voltage drop across any resisistor in a series circuit is determined by using the voltage divider rule, namely

$$V_x = \frac{R_x}{R_T} E \qquad (2\text{-}3)$$

CALCULATIONS

1. Refer to the the circuit of Figure 2-1. Calculate R_T, I, and the voltage across each resistor. Enter the results in Table 2-2. Show the correct units for each entry.

Resistor	Voltage
$R_1 = 47\ \Omega$	
$R_2 = 100\ \Omega$	
$R_3 = 270\ \Omega$	
$R_4 = 330\ \Omega$	
$R_5 = 470\ \Omega$	
R_T	
I	

Table 2-2

MEASUREMENTS

2. Connect the resistors as shown in the network of Figure 2-2. Use the DMM (ohmmeter) to measure the resistance across the open terminals. Enter the result below. Compare your measurement to the theoretical calculation recorded in Table 2-2. You should observe only a small discrepancy (no greater than the percent tolerance of the resistors.)

R_T	

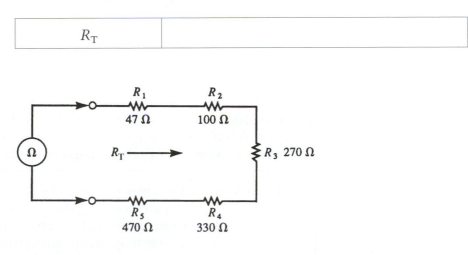

Figure 2-2 Series resistance

3. Connect the voltage source into the circuit as shown in Figure 2-3. With a DMM (voltmeter) connected across the voltage source, adjust the voltage for exactly 10 V.

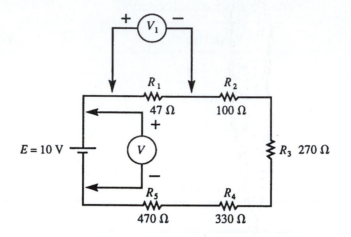

Figure 2-3 Measuring voltage in a series circuit

4. Disconnect the voltmeter from the voltage source and successively place it across each of the resistors. Measure the voltage across each resistor in the circuit and record your results in Table 2-3.

	Voltage
V_1	
V_2	
V_3	
V_4	
V_5	

Table 2-3

5. Turn off the voltage source and switch the DMM from the voltage range to the current range. Disconnect the circuit and insert the ammeter in series with the circuit as shown in Figure 2-4.

Points to Note:	The ammeter must be placed in series with the circuit, allowing the circuit current to pass through the meter. It must never be connected across an element, since this will result in a *short circuit* and may damage the meter or the circuit.

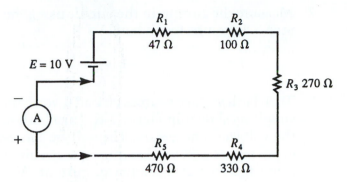

Figure 2-4   Current measurement

Turn the voltage source back on and record the circuit current in the space provided here. If all resistors have a 5% tolerance, you should measure a current that is within 5% of the calculated value of Table 2-2.

I	

6. Obtain a second voltage source or use the second supply of a dual power supply. Modify the series circuit by placing the second source in a *series-aiding connection* with the first voltage source as illustrated in Figure 2-5. Adjust the second supply for 5 V. (The first supply is still 10 V.) Notice that in the series-aiding connection, the positive terminal of the second source is connected to the negative terminal of the first source. Measure and record the voltage drop across each resistor.

	Voltage
V_1	
V_2	
V_3	
V_4	
V_5	

Table 2-4

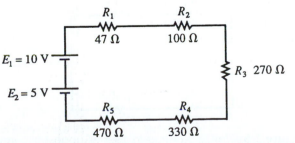

Figure 2-5   Voltage sources in a series-aiding connection

7. Measure the current in the circuit using the method described in Step 5. Enter the result here.

I	

8. Turn both voltage sources off and reverse the terminals of the 5-V supply as shown in Figure 2-6. Turn both supplies on and adjust the voltages if necessary. The voltage sources are now in a *series-opposing connection*. Examine the circuit of Figure 2-6 and determine the correct direction of current. Measure and record the voltage drop across each resistor.

	Voltage
V_1	
V_2	
V_3	
V_4	
V_5	

Table 2-5

9. Measure the current in the circuit using the method described in Step 5. Enter the result here.

I	

Ground

Electrical and electronic circuits are often connected to *ground*, meaning that this part of the circuit is at the same potential as the ground connection on a three-terminal plug. Since all grounds are connected in a building's electrical panel (as well as the water pipes), special precautions may need to be followed when using instru-

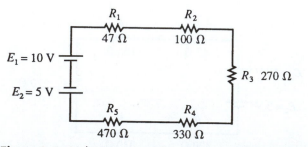

Figure 2-6   Voltage sources in a series-opposing connection

ments that are also grounded. Most dc voltage sources have a separate ground terminal at the front of the instrument, which permit us to connect a circuit to ground.

10. Refer to Figure 2-7(a). Connect the common terminals between the voltage sources to the ground terminal by using either a jumper wire or the grounding strip provided with the voltage source(s). Figure 2-7(b) indicates an alternate way of representing the voltage sources as *point sources*. Point sources simply indicate the potential at a point with respect to the reference (or in this case, ground). The ground symbol may or may not be shown. Connect the common (–) terminal of the DMM (voltmeter) to the ground point. All further voltage measurements are taken with respect to this point. Connect the voltage (+) terminal of the voltmeter to measure E_1, E_2, V_a, V_b, V_c, V_d. Record your results in Table 2-6.

	Voltage
E_1	
E_2	
V_a	
V_b	
V_c	
V_d	

Table 2-6

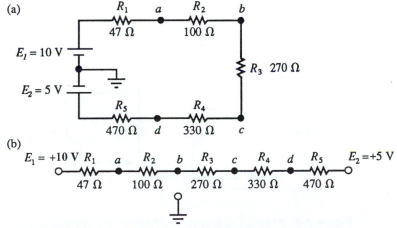

Figure 2-7 Ground connections and point sources

Internal Resistance of Voltage Sources

All voltage sources have some internal resistance, which tends to reduce the voltage between the terminals of the source when the
source is under load. The internal resistance of the voltage sources
used up to now have very small internal resistance (typically less
than 1 Ω). Other sources such as nickel-cadmium and alkaline batteries have relatively large internal resistance. This means that as a
circuit requires more current, the voltage across the terminals of the
battery will decrease. The magnitude of the internal resistance determines the maximum amount of usable current that a voltage source
can provide to a circuit.

11. Use a DMM to measure the voltage across the terminals of a 1.5-
    V D-cell alkaline battery and a 9-V alkaline battery. Enter the re-
    sults here.

E(1.5-V cell)	
E(9-V cell)	

12. Using each of the resistors from the previous part of the lab, con-
    struct the circuit shown in Figure 4-8. Measure and record the
    voltage across the terminals of the batteries for each of the resis-
    tors in Table 2-7.

CAUTION:	When connecting some of the resistors across the 9-V battery, the 1/4-W power rating will be exceeded. You will need to use 2-W resistors to permit voltage measurements with the 47-Ω, 100-Ω and 270-Ω resistors. Since the resistors may get quite hot, extra precaution should be observed when handling them.

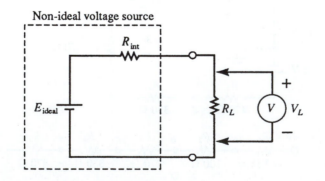

Figure 2-8   Measuring internal resistance of a voltage source

	V_L (1.5-V cell)	V_L (9-V cell)
$R_L = 470\ \Omega$		
$R_L = 330\ \Omega$		
$R_L = 270\ \Omega$		
$R_L = 100\ \Omega$		
$R_L = 47\ \Omega$		

Table 2-7

CONCLUSIONS

13. Compare the measured resistance of Step 2 to the theoretical resistance recorded in Table 2-2. Determine the percentage variation as shown.

R_T (theoretical) = _____ R_T (measured) = _____

$$\text{percent variation} = \frac{\text{Measurement–Theoretical}}{\text{Theoretical}} \times 100\% \quad (2\text{-}4)$$

percent variation = _____

14. Compare the measured voltage drops of Table 2-3 to the theoretical values recorded in Table 2-2. Indicate which values (if any) have a variation more than the resistor tolerance. Offer an explanation.

_____

_____

15. Examine the measured current of Step 5 and compare it to the theoretical value recorded in Table 2-2. Determine the percentage variation.

percent variation = _____

16. Determine the summation of the voltage drops recorded in Table 2-4. Compare this value to the summation of voltage rises. Is Kirchhoff's voltage law satisfied?

$\Sigma V =$ _____ $\Sigma E =$ _____

17. Calculate the theoretical current for the circuit of Figure 2-5. Compare this value to the measure value and determine the percent variation.

$I_{\text{theoretical}}$ = _____   percent variation = _____

18. Determine the summation of the voltage drops recorded in Table 2-5. Compare this value to the summation of voltage rises. Is Kirchhoff's voltage law satisfied?

ΣV = _____ ΣE = _____

_____

_____

19. Calculate the theoretical current for the circuit of Figure 4-6. Compare this value to the measure value and determine the percent variation.

$I_{\text{theoretical}}$ = _____ percent variation = _____

20. Calculate the voltages V_{ab}, V_{bc}, and V_{cd} from your measurements in Table 2-6. Compare your calculations to the corresponding measurements recorded in Table 4-5. Notice the direct correlation.

$V_{ab} = V_a - V_b$ = _____ V_2 = _____

V_{bc} = _____ V_3 = _____

V_{cd} = _____ V_4 = _____

21. For each load resistor, evaluate the internal resistance of the 1.5-V battery by applying Ohm's law as shown below. Enter your results in Table 2-8. Use these results to determine the average value of internal resistance.

$$I = \frac{V_L}{R_L} \tag{2-5}$$

$$R_{\text{int}} = \frac{E_{\text{ideal}} - V_L}{I} = \left(\frac{E_{\text{ideal}} - V_L}{V_L}\right) R_L \tag{2-6}$$

R_L	R_{int}
470 Ω	
330 Ω	
270 Ω	
100 Ω	
47 Ω	
Average value of R_{int}	

Table 2-8

22. For each load resistor, use equations (2-5) and (2-6) to evaluate the internal resistance of the 9-V battery. Enter your results in Table 2-9. Use these results to determine the average value of internal resistance.

R_L	R_{int}
470 Ω	
330 Ω	
270 Ω	
100 Ω	
47 Ω	
Average value of R_{int}	

Table 2-9

FOR FURTHER INVESTIGATION AND DISCUSSION

Use MultiSIM or PSpice to simulate the circuit of Figure 2-3. Determine the voltage drop across each resistor in the circuit and compare your results to those of Table 2-3. (Note: In order simulate the circuit, you will need to select a suitable reference point.)

LAB 3

Parallel dc Circuits

OBJECTIVES

After completing this lab, you will be able to
* assemble a parallel circuit consisting of a voltage source and several resistors,
* measure voltage and current in a parallel circuit,
* compare measured values to theoretical calculations and verify Kirchhoff's current law.
*

EQUIPMENT REQUIRED

☐ Digital multimeter (DMM)
☐ dc power supply
  Note: Record this equipment in Table 3-1.

COMPONENTS

☐ Resistors: 470-Ω, 680-Ω, 1-kΩ, 2.2-kΩ, 4.7-kΩ (1/4-W, 5%)

EQUIPMENT USED

Instrument	Manufacturer/Model No.	Serial No.
DMM		
dc Supply		

Table 3-1

REFERENCE

Section 6.1   PARALLEL CIRCUITS
Section 6.2   KIRCHHOFF'S CURRENT LAW
Section 6.3   RESISTORS IN PARALLEL
Section 6.5   CURRENT DIVIDER RULE
Section 6.6   ANALYSIS OF PARALLEL CIRCUITS

DISCUSSION

Two elements are said to be in a *parallel* connection if they have exactly two nodes in common. Each element in a parallel circuit, as shown in Figure 3-1, has the same voltage across it.

The equivalent conductance of n resistors in parallel, is determined as the summation of conductance

$$G_T = G_1 + G_2 + ... + G_n \qquad (3\text{-}1)$$

where the conductance G, of each resistor is found as the reciprocal of resistance

$$G_x = \frac{1}{R_x} \qquad (3\text{-}2)$$

The total resistance of n resistors in parallel is then found as

$$R_T = \frac{1}{G_T} \qquad (3\text{-}3)$$

When a parallel network of resistors is connected in parallel with a voltage source, the current through the voltage source is determined as

$$I = \frac{E}{R_T} \qquad (3\text{-}4)$$

The current through any resisistor in a parallel circuit is calculated using Ohm's law or the current divider rule, namely

$$I_x = \frac{E}{R_x} = \frac{R_T}{R_x}I \qquad (3\text{-}5)$$

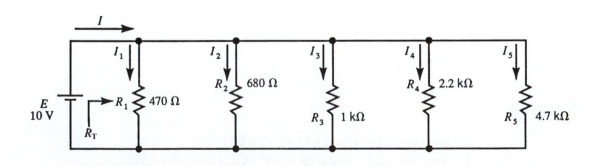

Figure 3-1   Parallel circuit

CALCULATIONS

1. Refer to the the circuit of Figure 3-1. Calculate R_T, I, I_1, I_2, I_3, I_4, and I_5. Enter the results in Table 3-2. Show the correct units for each entry.

I_1	
I_2	
I_3	
I_4	
I_5	

R_T	
I	

Table 3-2

MEASUREMENTS

2. Connect the resistors as shown in the network of Figure 3-2. Use the DMM (ohmmeter) to measure the resistance across the open terminals. Enter the result here. Compare your measurement to the theoretical calculation in Table 3-2. You should observe only a small discrepancy.

R_T	

3. Connect the voltage source to the circuit as shown in Figure 3-3. With a DMM (voltmeter) connected across the voltage source, adjust the voltage for exactly 10 V.

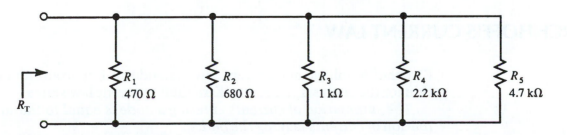

Figure 3-2 Parallel resistance

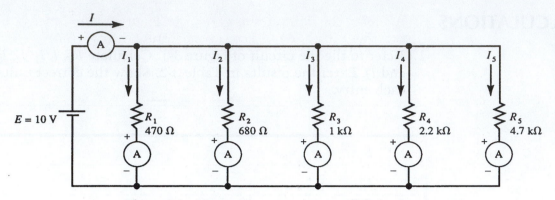

Figure 3-3   Measuring current in a parallel circuit

4. Disconnect the voltmeter from the voltage source. Turn the power supply off, without disturbing the voltage setting on the power supply. Set the DMM to measure current. Sucessively disconnect each branch of the circuit and insert the ammeter into the branch as illustrated in Figure 3-3. Measure the current through the voltage source and through each resistor in the circuit. **Ensure that each branch is reconnected after the ammeter is removed**. Record your results in Table 3-3.

	Current
I	
I_1	
I_2	
I_3	
I_4	
I_5	

Table 3-3

KIRCHHOFF'S CURRENT LAW

Kirchhoff's voltage and current laws provide an important foundation for the analysis of circuits. Kirchhoff's current law states:

The summation of currents entering a node is equal to the summation of currents leaving the node.

We now examine how Kirchhoff's current law can be verified in a laboratory.

5. Relocate the ammeter as shown in Figure 3-4 and measure the currents I_6, I_7, and I_8. Record your results in Table 3-4.

Figure 3-4   Verifying Kirchhoff's Current Law

	Current		
I_6			
I_7			
I_8			

Table 3-4

CONCLUSIONS

6. Compare the measured resistance of Step 2 to the theoretical resistance recorded in Table 3-2. Determine the percentage variation as shown.

R_T (theoretical) = _____ R_T (measured) = _____

$$\text{percent variation} = \frac{\text{Measurement--Theoretical}}{\text{Theoretical}} \times 100\% \quad (3\text{-}6)$$

percent variation = _____

7. Compare the measured currents of Step 4 to the theoretical values recorded in Table 3-2. Indicate which values (if any) have a variation more than the resistor tolerance. Offer an explanation.

_____

_____

8. Refer to the data of Table 3-3 and Table 3-4.
   a. Compare current I to the the summation $I_1 + I_2 + I_3 + I_4 + I_5$.

   $I = $ _____ $I_1 + I_2 + I_3 + I_4 + I_5 = $ _____

   b. Compare current I_6 to the the summation $I_2 + I_3 + I_4 + I_5$.

   $I_6 = $ _____ $I_2 + I_3 + I_4 + I_5 = $ _____

   c. Compare current I_7 to the the summation $I_3 + I_4 + I_5$.

   $I_7 = $ _____ $I_3 + I_4 + I_5 = $ _____

   d. Compare current I_8 to the the summation $I_4 + I_5$.

   $I_8 = $ _____ $I_4 + I_5 = $ _____

   e. Do the above calculations and measurements verify Kirchhoff's current law? Explain your answer.

   _____

   _____

FOR FURTHER INVESTIGATION AND DISCUSSION

Use MultiSIM or PSpice to simulate the circuit of Figure 3-1. Determine the current through each resistor in the circuit. Compare your results to those of Table 3-2.

Series-Parallel dc Circuits

OBJECTIVES

After completing this lab, you will be able to
- assemble a series-parallel circuit consisting of a voltage source and several resistors,
- measure voltage and current in a series-parallel circuit,
- compare measured values to theoretical calculations and verify Kirchhoff's current and voltage laws,
- assemble a zener diode regulator circuit and measure voltages and currents to verify that Kirchhoff's voltage and current laws apply,
- calculate power and verify the law of conservation of energy.

EQUIPMENT REQUIRED

☐ Digital multimeter (DMM)
☐ dc power supply
Note: Record this equipment in Table 4-1.

COMPONENTS

☐ Resistors: 330-Ω, 470-Ω, 680-Ω, 1-kΩ, 2.2-kΩ, 4.7-kΩ (1/4-W, 5%)
☐ Zener diode: 1N4734A (1-W, 5.6-V5%)

EQUIPMENT USED

Instrument	Manufacturer/Model No.	Serial No.
DMM		
dc Supply		

Table 4-1

TEXT REFERENCE

Section 7.1   THE SERIES-PARALLEL NETWORK
Section 7.2   ANALYSIS OF SERIES-PARALLEL CIRCUITS
Section 7.3   APPLICATIONS OF SERIES-PARALLEL CIRCUITS

DISCUSSION

Regardless of the complexity of a circuit, the basic laws of circuit analysis always apply. While Ohm's law and Kirchhoff's voltage and current laws are used to analyze simple series and parallel circuits, these same laws may be applied to analyze even the most complicated circuit. The following rules apply to all circuits.

The same current occurs through all series elements.

The same voltage appears across all parallel elements.

CALCULATIONS

1. Refer to the the circuit of Figure 4-1. Calculate the total resistance, R_T, seen by the voltage source. Calculate the current, I. Solve for all resistor currents, voltages, and powers. Enter the results in Table 4-2. Show the correct units for all entries.

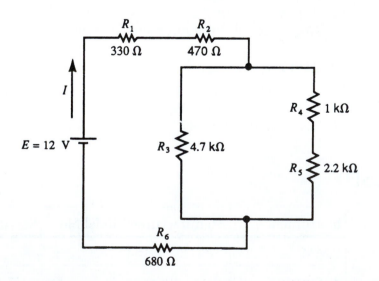

Figure 4-1   Series-parallel circuit

	Current	Voltage	Power
R_1			
R_2			
R_3			
R_4			
R_5			
R_6			

R_T	
I	

Table 4-2   Series-parallel resistance

MEASUREMENTS

2. Connect the resistors as shown in the network of Figure 4-2. Use the DMM (ohmmeter) to measure the resistance across the open terminals. Enter the result here. Compare your measurement to the theoretical calculation in Table 4-2. You should observe only a small discrepancy.

R_T	

3. Connect the voltage source to the circuit as shown in Figure 4-1. With a DMM (voltmeter) connected across the voltage source, adjust the voltage for exactly 12 V.

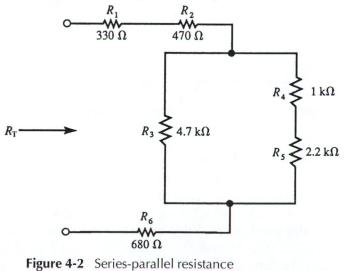

Figure 4-2   Series-parallel resistance

4. Disconnect the voltmeter from the voltage source. Measure the voltage across each resistor in the circuit and record the results in Table 4-3.
5. Set the DMM to measure the currents as illustrated in Figure 4-3. **Ensure that each branch is reconnected after the ammeter is removed.** Record your measurements in Table 4-3.

	Current	Voltage
R_1		
R_2		
R_3		
R_4		
R_5		
R_6		

I	

Table 4-3

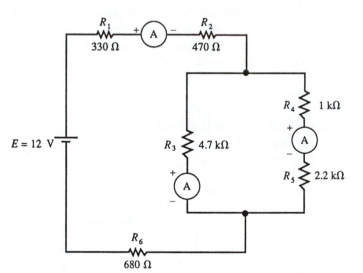

Figure 4-3   Measuring current in a series-parallel circuit

Zener Diode Circuit

In this part of the lab, we apply the principles of circuit analysis to examine the operation of a more complicated circuit. Here we use a *zener diode*, which is a two-terminal semiconductor device normally used as a *voltage regulator* to maintain a constant voltage between two terminals. When the zener diode is placed across a component which has a voltage greater than the *break-over* (zener) voltage of the diode, current through the zener diode forces the voltage across the component to decrease. While most other diodes permit current in only the forward direction (in the direction of the arrow in the diode symbol), the zener diode can conduct in either direction. When used as a voltage regulator, the zener diode is operated in the reverse-biased conditon. This means that when the zener diode is in its *breakover region*, current is against the arrow of the diode symbol.

6. Assemble the circuit shown in Figure 4-4, temporarily omitting the zener diode. Adjust the voltage source for 12 V.
7. Measure the voltages across R$_1$ and R$_2$ and record the values here.

V_1	
V_2	

8. Insert the zener diode into the circuit. Use the DMM (voltmeter) to measure voltages, V_1 and V_2 in the circuit of Figure 4-4. Convert the DMM to measure current and correctly measure currents I_1, I_2, and I_Z. Make sure that you turn off the voltage supply before disconnecting the circuit to insert the ammeter. In the circuit of Figure 4-4, show where you placed the ammeters to measure the currents. Record all measurements in Table 4-4.

Current	Voltage
$I_1 =$	$V_1 =$
$I_2 =$	$V_2 =$
$I_Z =$	

Table 4-4

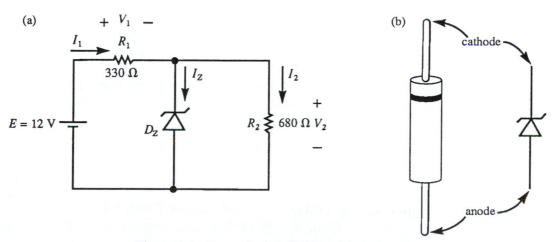

Figure 4-4   Zener diode voltage regulator circuit

CONCLUSIONS

9. Compare the measured resistance of Step 2 to the theoretical value recorded in Table 4-2. Determine the percent variation.

R_T (theoretical) = _____ R_T (measured) = _____

percent variation = _____

10. Compare the measured voltage drops in Table 4-3 to the theoretical values recorded in Table 4-2. Indicate which values (if any) have a variation more than the resistor tolerance. Offer an explanation.

_____

_____

11. Compare the measured currents in Table 4-3 to the theoretical values recorded in Table 4-2. Indicate which values (if any) have a variation more than the resistor tolerance. Offer an explanation.

_____

_____

12. Refer to the data of Table 4-3.

    a. Compare current I to the the summation $I_3 + I_4$

    I = _____ $I_3 + I_4$ = _____

    b. Do the above calculations and measurements satisfy Kirchhoff's current law? Explain your answer.

13. a. Use the data of Table 4-3 to calculate the power, P_T delivered to the circuit by the voltage source. Enter the result here.

P_T	

    b. Use the voltages and currents of Table 4-3 to determine the power dissipated by each resistor in the circuit. Enter your results in Table 4-5.

	Power
R_1	
R_2	
R_3	
R_4	
R_5	
R_6	

Table 4-5

   c. Compare the total power dissipated by the resistors to the total power delivered by the voltage source. Is energy conserved?

_____

_____

14. Refer to the circuit of Figure 4-4. Compare the voltage, V_2, with the zener diode removed from the circuit, to the voltage when the diode is in the circuit. How do they compare? Explain briefly why this occurred.

_____

_____

15. Refer to the data of Table 4-4.
   a. Compare current I_1 to the the summation $I_2 + I_Z$.

$I_1 = $ _____ $I_2 + I_Z = $ _____

   b. Do the calculations and measurements in part a) satisfy Kirchhoff's current law? Explain your answer.

_____

_____

16. a. Use the data in Table 4-4 to calculate the total power, P_T delivered to the circuit by the voltage source. Enter the result here.

P_T	

b. Use the voltages and currents of Table 4-4 to determine the power dissipated by the zener diode and by each resistor in the circuit. Enter your results in Table 4-6.

	Power
R_1	
R_2	
D_Z	

Table 4-6

c. Compare the total power dissipated by the resistors to the total power delivered by the voltage source. Is energy conserved?

_____

_____

PROBLEMS

17. Refer to the circuit of Figure 4-4. Assume that the zener diode has a break-over voltage of 4.3 V.
    a. Calculate the voltages V_1 and V_2
    b. Determine the currents I_1, I_2 and I_Z
    c. Solve for the powers dissipated by R_1, R_2, and D_Z.
    d. Show that the total power dissipated is equal to the power delivered by the voltage source.

Potentiometers and Rheostats

OBJECTIVES

After completing this lab, you will be able to
- demonstrate the use a variable resistor as a potentiometer to control the voltage applied to a load,
- demonstrate the use a variable resistor as a rheostat to control the current applied to a load,
- measure how the value of a load's resistance affects the voltage across a potentiometer.

EQUIPMENT REQUIRED

☐ Digital multimeter (DMM)
☐ dc power supply
   Note: Record this equipment in Table 5-1.

COMPONENTS

☐ Resistors: 5.6-kΩ, 3.3-kΩ, 330-kΩ (1/4-W, 5%)
                10-kΩ variable resistor

EQUIPMENT USED

Instrument	Manufacturer/Model No.	Serial No.
DMM		
dc Supply		

Table 5-1

TEXT REFERENCE

Section 3.5   TYPES OF RESISTORS
Section 7.4   POTENTIOMETERS

DISCUSSION

Variable resistors are used extensively in electrical and electronic circuits to control the voltage and current in circuits. When a variable resistor is used to control the voltage (as in the volume control of an amplifier) it is called a *potentiometer*. If the same resistor is used to control the amount of current through a circuit (such as in a light dimmer) it is called a *rheostat*.

Refer to the series circuit of Figure 5-1. The current through this circuit is constant regardless of the location of the wiper arm (terminal b) of the variable resistor. If the wiper arm is moved so that it is at the bottom of the resistor, the resistance between terminals b and c is zero. This results in the voltage, V_L being zero volts. If, however, the wiper arm is moved so that it is at the top of the resistor, the resistance between terminals b and c will be at a maximum, resulting in a maximum voltage, V_L, appearing between the terminals.

CALCULATIONS

1. Determine the range of the output voltage, V_L for the circuit of Figure 5-1. Calculate the range of output voltage if a 330-kΩ resistor is connected across the output terminals of the circuit. Recalculate the range of output voltage if a 3.3-kΩ resistor is connected across the output terminals of the circuit. Record your results in Table 5-2.

	$V_L(\text{min})$	$V_L(\text{max})$
$R_L = \infty$ (open)		
$R_L = 330 \text{ k}\Omega$		
$R_L = 3.3 \text{ k}\Omega$		

Table 5-2

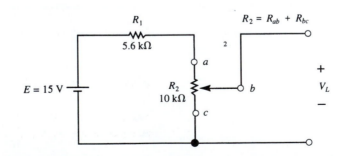

Figure 5-1   Variable resistor used as a potentiometer

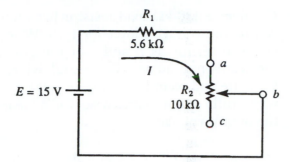

Figure 5-2 Variable resistor used as a rheostat

2. The 10-kΩ resistor in the circuit of Figure 5-1 is easily converted from a potentiometer into a rheostat. Figure 5-2 shows how the variable resistor is used as a rheostat. When the wiper arm is moved so that it is at the top of the resistor, the resistance of the rheostat will be at its minimum value. This means that maximum current will occur in the circuit. If the wiper is moved to the bottom of the resistor, the resistance of the rheostat will be at its maximum value, resulting in the least amount of current.

    Calculate the range of current for the circuit of Figure 5-2.

$I(\text{min})$	$I(\text{max})$

MEASUREMENTS

3. Assemble the circuit shown in Figure 5-1. Adjust the voltage source for 15 V. With a voltmeter connected between terminals b and c of the potentiometer, use a small screwdriver to adjust the central wiper fully clockwise (CW). Measure and record the output voltage. Now adjust the central wiper fully counterclockwise (CCW). Again measure and record the output voltage.

$V_L(\text{CW})$	$V_L(\text{CCW})$

4. Adjust the potentiometer to obtain an output voltage of 3.0 V. Disconnect the potentiometer from the circuit, being careful not to readjust the potentiometer setting. Measure and record the resistance between terminals b and c of the potentiometer.

R_{bc}	

5. Connect a 330-kΩ load resistor between the output terminals. Adjust the voltage source for 15 V. With a voltmeter connected between terminals b and c of the potentiometer, use a small screwdriver to adjust the central wiper fully clockwise (CW). Measure and record the output voltage. Now adjust the central wiper fully counterclockwise (CCW). Again measure and record the output voltage.

V_L(CW)	V_L(CCW)

6. Adjust the potentiometer to obtain an output voltage of 3.0 V. Disconnect the potentiometer from the circuit, being careful not to readjust the potentiometer setting. Measure and record the resistance between terminals b and c of the potentiometer.

R_{bc}	

7. Connect a 3.3-kΩ load resistor between the output terminals. Adjust the voltage source for 15 V. With a voltmeter connected between terminals b and c of the potentiometer, use a small screwdriver to adjust the central wiper fully clockwise (CW). Measure and record the output voltage. Now adjust the central wiper fully counterclockwise (CCW). Again measure and record the output voltage.

V_L(CW)	V_L(CCW

8. Adjust the potentiometer to obtain an output voltage of 3.0 V. Disconnect the potentiometer from the circuit, being careful not to readjust the potentiometer setting. Measure and record the resistance between terminals b and c of the potentiometer.

R_{bc}	

9. Construct the circuit of Figure 5-2. Place a DMM ammeter into the circuit to measure the current. Use a small screwdriver to adjust the central wiper fully clockwise (CW). Measure and record the circuit current. Now adjust the central wiper fully counterclockwise (CCW). Again measure and record the current.

I(CW)	I(CCW)

10. Adjust the rheostat so that the measured current is exactly equal to $I_{max}/2$. Remove the rheostat from the circuit, being careful not to readjust the potentiometer setting. Measure and record the resistance between terminals a and b of the potentiometer. The resistance should be exactly equal to R_1.

R_{ab}	

CONCLUSIONS

11. Compare the measurements of Step 3 to the theoretical maximum and minimum values of voltage recorded in Table 5-2 for $R_L = \infty$ (open).

_____

_____

_____

_____

12. Compare the measurements of Step 5 to the theoretical maximum and minimum values of voltage recorded in Table 5-2 for $R_L = 330\ k\Omega$.

_____

_____

_____

_____

13. Use the measured resistance, R_{bc} of Step 6 to determine the theoretical voltage which would appear across the load, $R_L = 330\ k\Omega$ in the equivalent circuit of Figure 5-3.

V_L	

Compare the above value to the measured load voltage, V_L in Step 6.

_____

14. Compare the measurements of Step 7 to the theoretical maximum and minimum values of voltage recorded in Table 5-2 for $R_L = 3.3 \text{ k}\Omega$.

_____

_____

15. Use the measured resistance, R_{bc} of Step 8 to determine the theoretical voltage which would appear across the load, $R_L = 3.3 \text{ k}\Omega$ in the equivalent circuit of Figure 5-3.

V_L	

   Compare the above value to the measured load voltage, V_L in Step 8.

_____

16. Compare the measurements of Step 9 to the theoretical maximum and minimum values of current recorded in Step 2.

_____

_____

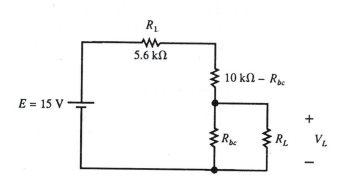

Figure 5-3

Superposition Theorem

OBJECTIVES

After completing this lab, you will be able to
- calculate currents and voltages in a dc circuit using the superposition theorem,
- measure voltage and current in a multi-source circuit,
- measure the effects of sucessively removing each voltage source from a circuit,
- calculate loop currents and node voltages using mesh analysis and nodal analysis,
- verify the superposition theorem as it applies to dc circuits and show that the results are consistent with results determined using mesh analysis and nodal analysis,

EQUIPMENT REQUIRED

☐ Digital multimeter (DMM)
☐ dc power supply (2)
Note: Record this equipment in Table 6-1.

COMPONENTS

☐ Resistors: 680-Ω, 1-kΩ, 3.3-kΩ (1/4-W, 5%)

EQUIPMENT USED

Instrument	Manufacturer/Model No.	Serial No.
DMM		
dc Supply		
dc Supply		

Table 6-1

TEXT REFERENCE

DISCUSSION

Mesh analysis allows us to find the loop currents for a circuit having any number of voltage or current sources. If a circuit contains current sources, these must first be converted to voltage sources.

Nodal analysis is the twin of mesh analysis in that it allows us to calculate nodal voltages of a circuit (with respect to a reference node). If a circuit contains voltage sources, it is necessary to first convert these to current sources.

Analyzing a circuit using mesh or nodal analysis usually requires solving several linear equations. The superposition theorem allows us to simplify the analysis of a multi-source circuit by considering only one source at a time.

The superposition theorem states:

The voltage across (or the current through) a resistor may be determined by finding the sum of the effects due to each independent source in the circuit.

In order to determine the effects due to one source, it is necessary to remove all other sources from the circuit. This is accomplished by replacing voltage sources with short circuits and by replacing current sources with open circuits.

CALCULATIONS

1. Use superposition to calculate currents I_1, I_2, and I_3 in the circuit of Figure 6-1. Record the results.

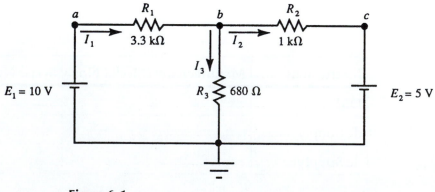

Figure 6-1

I_1	
I_2	
I_3	

2. Use superposition to calculate the currents I_1, I_2, and I_3 in the circuit of Figure 6-2. Record the results below.

I_1	
I_2	
I_3	

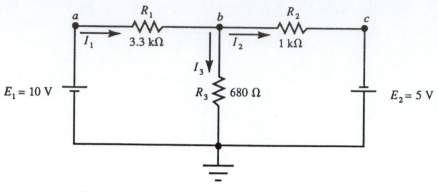

Figure 6-2

MEASUREMENTS

3. Assemble the circuit of Figure 6-1. Measure the voltages V_{ab}, V_b, and V_{bc}. Record your results in Table 6-2, showing the correct polarity for each measurement. Use these measurements and the resistor color codes to calculate the currents I_1, I_2, and I_3 for the circuit. Show the correct polarity for each current. (Use a negative sign to indicate that current is opposite to the indicated direction.)

V_{ab}	
V_b	
V_{bc}	
I_1	
I_2	
I_3	

Table 6-2

4. Assemble the circuit of Figure 6-2. Measure the voltages V_{ab}, V_b, and V_{bc}. Record your results in Table 6-3, showing the correct polarity for each measurement. Use these measurements and the resistor color codes to calculate the currents I_1, I_2, and I_3 for the circuit. Show the correct polarity for each current.

V_{ab}	
V_b	
V_{bc}	
I_1	
I_2	
I_3	

Table 6-3

5. Remove the voltage source E_2 from the circuit of Figure 6-1 and replace it with a short circuit. Measure the voltages V_{ab}, V_b, and V_{bc} and calculate the currents I_1, I_2, and I_3 due to the voltage source E_1. Record these results in Table 6-4.

$V_{ab(1)}$	
$V_{b(1)}$	
$V_{bc(1)}$	
$I_{1(1)}$	
$I_{2(1)}$	
$I_{3(1)}$	

Table 6-4

6. Remove voltage source E_1 from the circuit of Figure 6-1 and replace it with a short circuit. Measure voltages V_{ab}, V_b, and V_{bc} and use the results to calculate currents I_1, I_2, and I_3 due to the voltage source E_2. Record the results in Table 6-5.

$V_{ab(2)}$	
$V_{b(2)}$	
$V_{bc(2)}$	
$I_{1(2)}$	
$I_{2(2)}$	
$I_{3(2)}$	

Table 6-5

7. Reverse the polarity of voltage source E_2 as shown in the circuit of Figure 6-2. (Voltage source E_1 is still substituted by a short circuit.) Measure voltages V_{ab}, V_b, and V_{bc} and use the results to calculate currents I_1, I_2, and I_3 due to the voltage source E_2. Record the results in Table 6-6.

$V_{ab(3)}$	
$V_{b(3)}$	
$V_{bc(3)}$	
$I_{1(3)}$	
$I_{2(3)}$	
$I_{3(3)}$	

Table 6-6

CONCLUSIONS

8. Compare the measured currents of Table 6-2 to the theoretical currents calculated in Step 1. Determine the percent deviation for each of the currents.

I_1(theoretical) = _____ percent variation = _____

I_2(theoretical) = _____ percent variation = _____

I_3(theoretical) = _____ percent variation = _____

9. Compare the measured currents recorded in Table 6-3 to the theoretical currents calculated in Step 2. Determine the percent deviation for each of the currents.

I_1(theoretical) = _____ percent variation = _____

I_2(theoretical) = _____ percent variation = _____

I_3(theoretical) = _____ percent variation = _____

10. Combine the results of Table 6-4 and Table 6-5.

$$V_{ab} = V_{ab(1)} + V_{ab(2)} = \text{_____}$$

$$V_b = V_{b(1)} + V_{b(2)} = \text{_____}$$

$$V_{bc} = V_{bc(1)} + V_{bc(2)} = \text{_____}$$

$$I_1 = I_{1(1)} + I_{1(2)} = \text{_____}$$

$$I_2 = I_{2(1)} + I_{2(2)} = \text{_____}$$

$$I_3 = I_{3(1)} + I_{3(2)} = \text{_____}$$

Compare the previous results to the measurements recorded in Table 6-2. According to superposition, the results should be the same.

_____

_____

11. Combine the results of Table 6-4 and Table 6-6.

$$V_{ab} = V_{ab(1)} + V_{ab(3)} = \underline{\hspace{2cm}}$$

$$V_b = V_{b(1)} + V_{b(3)} = \underline{\hspace{2cm}}$$

$$V_{bc} = V_{bc(1)} + V_{bc(3)} = \underline{\hspace{2cm}}$$

$$I_1 = I_{1(1)} + I_{1(3)} = \underline{\hspace{2cm}}$$

$$I_2 = I_{2(1)} + I_{2(3)} = \underline{\hspace{2cm}}$$

$$I_3 = I_{3(1)} + I_{3(3)} = \underline{\hspace{2cm}}$$

Compare the above results to the measurements recorded in Table 6-3. According to superposition, the results should be the same.

_____

_____

PROBLEMS

12. Apply mesh analysis to calculate the loop currents in the circuit of Figure 6-1. Use the loop currents to determine currents I_1, I_2, and I_3.

I_1	
I_2	
I_3	

$I_{\text{loop 1}} =$ _____

$I_{\text{loop 2}} =$ _____

13. Apply mesh anaysis to calculate the loop currents in the circuit of Figure 6-2. Use the loop currents to determine currents I_1, I_2, and I_3.

$I_{\text{loop 1}} =$ _____

$I_{\text{loop 2}} =$ _____

I_1	
I_2	
I_3	

14. Use nodal analysis to calculate the node voltage V_b in the circuit of Figure 6-1. Your result should be very close to the measured value recorded in Table 6-2.

$V_b = $ _____

15. Use nodal analysis to calculate the node voltage V_b in the circuit of Figure 6-2. Your result should be very close to the measured value recorded in Table 6-3.

$\mathbf{V_b} = $ _____

Thévenin's and Norton's Theorems (dc)

OBJECTIVES

After completing this lab, you will be able to
- determine the Thévenin and Norton equivalent of a complex circuit,
- analyze a circuit using the Thévenin or Norton equivalent circuit,
- determine the value of load resistance needed to ensure maximum power transfer to the load,
- measure the Thévenin (or Norton) resistance of a circuit using an ohmmeter,
- measure the Thévenin voltage of a circuit using a voltmeter,
- measure the Norton current of a circuit using an ammeter.
- describe how maximum power is transferred to a load when the load resistance is equal to the Thévenin resistance ($R_L = R_{Th}$).

EQUIPMENT REQUIRED

☐ Digital multimeter (DMM)
☐ dc power supply
   Note: Record this equipment in Table 7-1.

COMPONENT:

☐ Resistors: 3.3-kΩ, 4.7-kΩ, 5.6-kΩ (1/4-W, 5% tolerance)
                  10-kΩ variable resistor

EQUIPMENT USED

Instrument	Manufacturer/Model No	Serial No.
DMM		
dc Supply		

Table 7-1

TEXT REFERENCE

Section 9-2   THÉVENIN'S THEOREM
Section 9.3   NORTON'S THEOREM
Section 9.4   MAXIMUM POWER TRANSFER THEOREM

DISCUSSION

Thévenin's theorem:
Any linear bilateral network may be reduced to a simplified two-terminal network consisting of a single voltage source, E_{Th}, in series with a single resistor, R_{Th}. Once the original network is simplified, any load connected to the output terminals will behave exactly as if the load were connected in series with E_{Th} and R_{Th}.

Norton's theorem:
Any linear bilateral network may be reduced to a simplified two-terminal network consisting of a single current source, I_N, in parallel with a single resistor, R_N. A Thévenin equivalent circuit is easily converted into a Norton equivalent by performing a source conversion as follows:

$$R_N = R_{Th} \tag{7-1}$$

$$I_N = \frac{E_{Th}}{R_{Th}} \tag{7-2}$$

When a load is connected across the output terminals, the circuit will behave exactly as if the load were connected in parallel with I_N and R_N.

Maximum power transfer theorem:
Maximum power will be delivered to the load resistance when the load resistance is equal to the Thévenin (or Norton) resistance.

CALCULATIONS

1. Determine the Thévenin equivalent of the circuit of Figure 7-1. Sketch the equivalent circuit in the space provided below.

2. Determine the Norton equivalent of the circuit of Figure 7-1. Sketch the equivalent circuit in the space provided below.

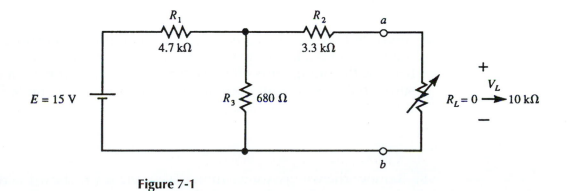

Figure 7-1

3. Calculate the minimum and the maximum voltage V_L which will appear across the load as R_L is varied between 0 and 10 kΩ. Enter the results below.

$V_{L(min)}$	
$V_{L(max)}$	

4. Calculate the minimum and the maximum load current I_L which will occur through the load as R_L is varied between 0 and 10 kΩ. Enter the results below.

$I_{L(min)}$	
$I_{L(max)}$	

5. Determine the value of load resistance for which maximum power will be transferred to the load.

MEASUREMENTS

6. Assemble the circuit of Figure 7-1, temporarily omitting the load resistor R_L. Insert the DMM voltmeter across terminals a and b and measure the open-circuit voltage. This is the Thévenin voltage E_{Th}. Record the measurement here.

E_{TH}	

7. Insert the DMM ammeter between terminals a and b. Ensure that the ammeter is adjusted to measure the expected Norton current. Because the ammeter is effectively a short circuit, you are measuring the short-circuit current between terminals a and b. Record the measurement here.

I_N	

8. Remove the voltage source from the circuit and replace it with a short circuit. Place the DMM ohmmeter across terminals a and

b and measure the resistance between these terminals. This is the Thévenin resistance R_{Th}. Record the result here.

R_{Th}	

9. Reconnect the voltage source into the circuit. Connect the variable 10-kΩ resistor as a rheostat and insert it as the load resistance R_L. Adjust the variable resistor between its minimum and maximum values. Measure and record the maximum and minimum output voltage V_L.

$V_{L\ (min)}$	
$V_{L\ (max)}$	

10. Place the DMM ammeter in series with the load resistor R_L. Adjust the variable resistor between its minimum and maximum values. Measure and record the minimum and maximum load current I_L.

$I_{L\ (min)}$	
$I_{L\ (max)}$	

11. Connect the DMM voltmeter across the load resistor. Adjust R_L until the output voltage is exactly half of the Thévenin voltage measured in Step 6. When $V_L = E_{Th}/2$, the load resistor is receiving the maximum amount of power from the circuit. Carefully remove R_L from the circuit, ensuring that the rheostat is not accidently readjusted. Use the DMM ohmmeter to measure the value of R_L. Record the measurement here.

R_L	

CONCLUSIONS

12. Compare the measured value of Thévenin voltage E_{Th} in Step 6 to the calculated value of Step 1. Determine the percent variation.

$E_{Th(theoretical)}$ = _____   percent variation = _____

13. Compare the measured value of Thévenin (or Norton) resistance R_{Th} in Step 8 to the calculated value of Step 1. Determine the percent variation.

$R_{Th(theoretical)}$ = _____   percent variation = _____

14. Compare the measured value of Norton current I_N in Step 7 to the calculated value of Step 2. Determine the percent variation.

$I_{N(theoretical)}$ = _____percent variation = _____

15. An alternate method of determining the Thévenin (Norton) resistance is by applying Ohm's law to the Thévenin voltage and Norton current. Calculate the value of Thévenin resistance using the measured values of Thévenin voltage and Norton current.

$$R_{Th} = R_N = \frac{E_{Th}}{I_N} \qquad (7\text{-}3)$$

R_{Th}	

    Compare this value to the actual measured value of Thévenin resistance of Step 8.

_____

_____

16. Compare the measured minimum and maximum voltage V_L to the calculated voltages determined in Step 3. Explain why there is a slight variation.

_____

_____

17. Compare the measured minimum and maximum load current I_L to the calculated currents determined in Step 4.

_____

18. When delivering maximum power to the load, how did the actual value of load resistance R_L compare the theoretical value calculated in Step 5.

_____

_____

Capacitors

OBJECTIVES

After completing this lab, you will be able to
- measure capacitance,
- verify capacitor relationships for series and parallel connections,
- verify that a capacitor behaves as an open circuit for steady state dc.

EQUIPMENT REQUIRED

☐ DMM
☐ Capacitance meter
☐ Variable dc power supply

COMPONENTS

☐ Capacitors: One each of 1 μF, 0.47 μF and 0.33 μF, non-electrolytic, 35 WVDC or greater
☐ Resistors: One each of 2.7-kΩ, 3.9-kΩ, and 10-kΩ, 1/4 W

EQUIPMENT USED

Instrument	Manufacturer/Model No.	Serial No.
DMM		
Capacitance meter[†]		
Power supply		

[†]Such as a DMM with capacitance measuring capability, an LCR meter (e.g., the BK Precision 878), a digital capacitance bridge, or an impedance bridge.

Table 8-1

TEXT REFERENCE

Section 10.1   CAPACITANCE
Section 10.7   CAPACITORS IN PARALLEL AND SERIES

DISCUSSION

> **Practical Note**
> The farad is a very large quantity; practical capacitors generally range in value from a few picofarads to a few hundred microfarads.

A *capacitor* is a charge storage device and its electrical property is called *capacitance*. The more charge that a capacitor can store for a given voltage, the larger its capacitance. Capacitance, charge and voltage are related by the equation

$$C = \frac{Q}{V} \qquad (8\text{-}1)$$

where C is capacitance in *farads*, Q is charge stored (in coulombs) and V is the capacitor terminal voltage (in volts). Because of its ability to store charge, a capacitor holds its voltage. That is, if you charge a capacitor, then disconnect the source, a voltage will remain on the capacitor for a considerable length of time. Dangerous voltages can be present on charged capacitors. For this reason, you should discharge capacitors before working with them.

For capacitors in parallel, the total capacitance is the sum of the individual capacitances. That is,

$$C_T = C_1 + C_2 + ... + C_N \qquad (8\text{-}2)$$

For capacitors in series, the total capacitance may be found from

$$\frac{1}{C_T} = \frac{1}{C_1} + \frac{1}{C_2} + ... + \frac{1}{C_N} \qquad (8\text{-}3)$$

Steady State Capacitor Currents and Voltages

Since capacitors consist of conducting plates separated by an insulator, there is no conductive path from terminal to terminal. Thus, when a capacitor is placed across a dc source, its steady state current is zero. This means that a capacitor in steady state looks like an open circuit to dc.

When connected in parallel, the voltage across all capacitors is the same. However, when connected in series, voltage divides in inverse proportion to the size of the capacitances: that is, the smaller the capacitance, the larger the voltage. For capacitors in series as in Figure 8-3(a), the steady dc voltages on capacitors are related by

$$V_x = \left(\frac{C_T}{C_x}\right)E, \quad V_1 = \left(\frac{C_2}{C_1}\right)V_2, \quad V_1 = \left(\frac{C_3}{C_1}\right)V_3 \qquad (8\text{-}4)$$

and so on. (This is the voltage divider rule for capacitance.)

Measuring Capacitance

Measuring capacitance with a modern capacitor tester is straightforward. The general procedure is

1. Short the capacitor's leads to discharge the capacitor. Remove the short.
2. Set the function selector of the tester to the appropriate capacitance range (if not autoranging), then connect the capacitor. Observe polarity markings on polarized capacitors if applicable. (This is not necessary on some testers.)
3. Read the capacitance value directly from the numeric readout.

With such testers, it takes only a few seconds to measure capacitance.

MEASUREMENTS

1. Carefully measure each capacitor and resistor and record their values in Tables 8-2 and 8-3.

	Nominal	Measured
C_1	1 µF	
C_2	0.47 µF	
C_3	0.33 µF	

Table 8-2

	Nominal	Measured
R_1	2.7 kΩ	
R_2	3.9 kΩ	
R_3	10 kΩ	

Table 8-3

2. a. Assemble the circuit of Figure 8-1(a), measure capacitance C_T and record in Table 8-4. Repeat for circuits (b), (c) and (d).
    b. Verify the values measured in (a) by analyzing each circuit using the measured capacitor values from Table 8-2. Record calculated values in Table 8-4. How do they compare to the measured results?

(a)

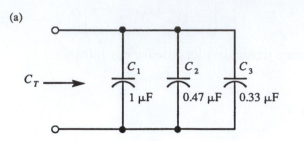

(b)

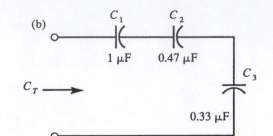

(c)

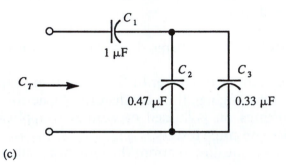

(d)

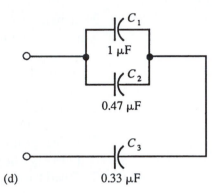

Figure 8-1 Circuits for Test 2

	Total Capitance C_T (µF)		
Circuit	Measured	Calculated	% of Difference
Figure 8-1(a)			
Figure 8-1(b)			
Figure 8-1(c)			
Figure 8-1(d)			

Table 8-4

3. a. Assemble the circuit of Figure 8-2. Set E = 18.0 V, measure V_1, V_2, and V_3 and record in Table 8-5.

	Measured	Computed
V_1		
V_2		
V_3		

Table 8-5

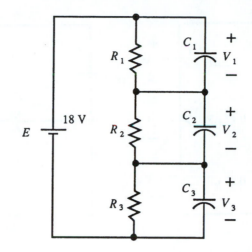

Figure 8-2 Circuit for Test 3

    b. Verify the results of Table 8-5 by analyzing the circuit of Figure 8-2 and calculating voltages. Why do Equations 8-4 not apply here? Why do the capacitors not load the circuit?

PROBLEMS

4. For the circuit of Figure 8-3(a), compute the voltage across each capacitor and record in Table 8-6. Repeat for Figure 8-3(b).

	Figure 8-3(a)	Figure 8-3(b)
V_1		
V_2		
V_3		

Table 8-6

5. For the circuit of Figure 8-4, compute the voltage across the capacitor.

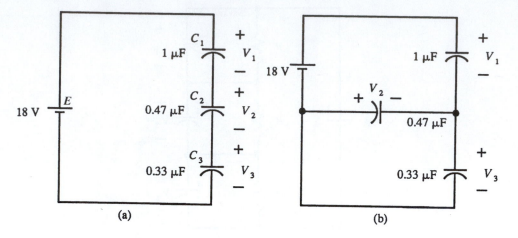

Figure 8-3 Circuits for Part 4

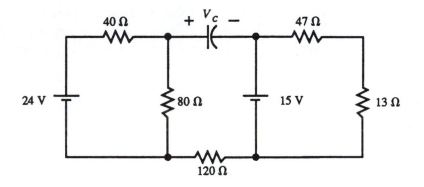

Figure 8-4 Circuit for Part 5

FOR FURTHER INVESTIGATION AND DISCUSSION

Capacitors can have quite loose tolerances—that is, the actual capacitance of a capacitor may differ by as much as ±10% to ±20% from its nominal value (i.e., its body marking) depending on the type of capacitor used. Consider two capacitors, $C_1 = 1\ \mu F \pm 10\%$ and $C_2 = 0.47\ \mu F \pm 20\%$. Write a short discussion paper on the impact of tolerances. As the basis for your discussion, (a) place the capacitors in parallel and compute the nominal capacitance C_T and the minimum and maximum value that C_T will have if both capacitors are at their worst-case tolerance limits. (b) Repeat if the capacitors are in series. (c) With this as background, show how tolerance errors can impact total capacitance of a more complex circuit. For this last part, consider Figure 8-1(c). Assume that C_3 has a tolerance of ±10%.

NAME _____

DATE _____

CLASS _____

Capacitor Charging and Discharging

OBJECTIVES

After completing this lab, you will be able to
- measure capacitor charge and discharge times,
- confirm the voltage/current direction convention for capacitors,
- confirm the Thévenin method of analysis for capacitive charging and discharging,

EQUIPMENT REQUIRED

☐ DMM, VOM
☐ Variable dc power supply
☐ Function generator (optional)
☐ Oscilloscope (optional)

COMPONENTS

☐ Capacitors: 470-μF (electrolytic), 0.01-μF (non-electrolytic)
☐ Resistors: 10-kΩ, 20-kΩ, 39-kΩ, 47-kΩ, 1/4-W
☐ Switch: Single pole, double throw
☐ Stopwatch

Note to the Instructor

Part D of this lab may be run as an instructor demo if your students have not yet had instruction on the oscilloscope.

EQUIPMENT USED

Instrument	Manufacturer/Model No.	Serial No.
DMM		
Power supply		
Function generator (optional)		
Oscilloscope (optional)		

Table 9-1

TEXT REFERENCE

Section 10.8   CAPACITOR CURRENT AND VOLTAGE
Section 11.1   INTRODUCTION
Section 11.2   CAPACITOR CHARGING EQUATIONS
Section 11.4   CAPACITOR DISCHARGING EQUATIONS
Section 11.5   MORE COMPLEX CIRCUITS
Section 11.8   TRANSIENT ANALYSIS USING COMPUTERS

DISCUSSION

Capacitor charging and discharging may be studied using the circuit of Figure 9-1. When the switch is in position 1, the capacitor charges at a rate determined by its capacitance and the resistance through which it charges; when the switch is in position 2, it discharges at a rate determined by its capacitance and the resistance through which it discharges. This phenomenon of charging and discharging is important as it affects the operation of many circuits.

> **Caution**
> In this lab, you use electrolytic capacitors. Electrolytics are polarized and must be used with their + lead connected to the positive side of the circuit and their − lead to the negative side. An incorrectly connected electrolytic may explode.

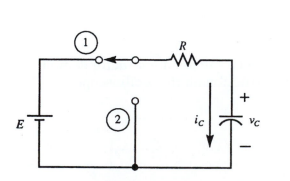

Figure 9-1   Circuit for studying capacitor charging and discharging

Voltage and Current During Charging

Consider Figure 9-2. At the instant the switch is moved to the charge position, current jumps from zero to E/R (since the capacitor looks like a short circuit at this instant). As the capacitor voltage approaches full source voltage, current approaches zero (since the capacitor looks like an open circuit to dc). During charging, voltage and current are given by

$$i_c = \frac{E}{R}e^{-t/RC} \qquad (9\text{-}1)$$

$$v_c = E(1 - e^{-t/RC}) \qquad (9\text{-}2)$$

The product RC is referred to as the *time constant* and is given the symbol τ. Thus,

$$\tau = RC \qquad (9\text{-}3)$$

In one time constant, the capacitor voltage climbs to 63.2% of its final value while the current drops to 36.8% of its initial value. For all practical purposes, charging is complete in five time constants.

Voltage and Current During Discharging

The discharge curves are shown in Figure 9-3. When the switch is moved to position 2, the capacitor looks momentarily like a voltage

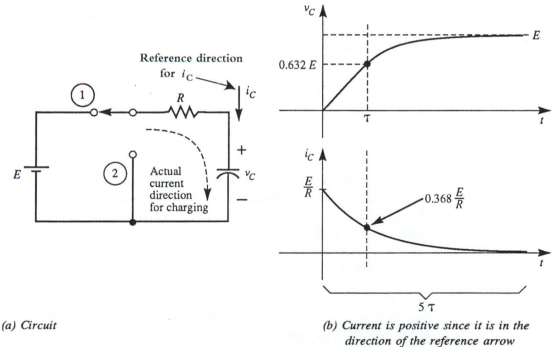

(a) Circuit

(b) Current is positive since it is in the direction of the reference arrow

Figure 9-2 Capacitor voltage and current during charging. Capacitor is initially uncharged

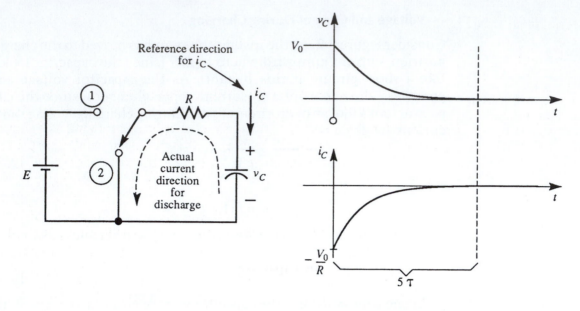

(a) Circuit

(b) Current is negative since it is opposite in direction to the reference arrow.

Figure 9-3   Capacitor voltage and current during discharging

source with value V_o where V_o is the voltage on the capacitor at the instant the switch is moved. (If the capacitor is fully charged, $V_o = E$.) Current jumps from zero to $-V_o/R$, then decays to zero. (It is negative since it is opposite in direction to the reference as indicated in Figure 9-3a.) Voltage decays from V_o to zero. Equations are

$$i_C = -\frac{V_o}{R}e^{-t/RC} \quad \text{or} \quad i_c = -\frac{E}{R}e^{-t/RC}$$

(9-4)

$$v_C = V_o e^{-t/RC} \quad \text{or} \quad v_c = E e^{-t/RC}$$

(9-5)

Discharging takes five time constants. Although for this particular circuit charge and discharge resistances are the same, in general they are different. For this latter case, charge and discharge time constants will be different.

A Final Note

To properly observe capacitor charging and discharging, you need an oscilloscope. Since you may not have studied the oscilloscope at this time, we will examine the basic ideas in other ways. However, at the end of the lab, we have included an instructor demonstration (Part D) using an oscilloscope. Here, you will be able to observe capacitor charging and discharging directly on the screen.

MEASUREMENTS

PART A: Charge and Discharge Times

For Part A, use a VOM to observe capacitor voltage and a stopwatch to time charging and discharging. For the large capacitance value required here, you need an electrolytic capacitor. Unfortunately, you will find that your results agree only moderately well with theory—see boxed Note 2. However, the results clearly verify the theory.

1. Assemble the circuit of Figure 9-1 with $R = 10$ kΩ, $C = 470$ µF and source voltage about 25 V. Place the switch in the charge position and wait for the capacitor to fully charge. Carefully adjust E until $V = 25$ V across the capacitor.

    a. Compute the time constant for the circuit.

    $$\tau_{computed} = RC = \underline{\hspace{3cm}}.$$

    b. Return the switch to the discharge position. Wait for the capacitor to fully discharge, then move the switch to the charge position and using the stopwatch, time how long it takes for the capacitor voltage to reach 15.8 V (i.e., 63.2% of its final voltage). This is the measured time constant.

    $$\tau_{measured} = \underline{\hspace{3cm}} \text{ (charging)}.$$

    c. Hold the switch in the charge position until the capacitor voltage stabilizes at 25 volts. Now move the switch to the discharge position and time how long it takes for the voltage to drop to 9.2 V (i.e., to 36.8% of its initial value).

    $$\tau_{measured} = \underline{\hspace{3cm}} \text{ (discharging)}.$$

2. Change R to 20 kΩ and repeat Test 1.

    a. $\tau_{computed} = \underline{\hspace{3cm}}$

    b. $\tau_{measured} = \underline{\hspace{3cm}}$ (charging)

    c. $\tau_{measured} = \underline{\hspace{3cm}}$ (discharging)

3. Describe how the results of Tests 1 and 2 confirm the theory.

    _____

    _____

PART B: Current Direction

To conform to the standard voltage/current convention, the + sign for voltage v_C must be at the tail of the current direction arrow i_C as indicated in Figure 9-1. This is obviously correct for charging as indicated in Figure 9-2. During discharge, the polarity of the voltage does not change. Thus, for discharging, the current direction arrow must also remain in the clockwise direction, even though we know that actual current is in the opposite direction as indicated in Figure 9-3. The interpretation is that for charging, current is in the same direction as the reference and hence, is positive, while for discharging, it is opposite to the reference and hence, is negative.

4. Add a DMM with autopolarity as in Figure 9-4. Connect the current input jack A to the positive side of the circuit as indicated. (Since the meter measures current *into* terminal A, this connection will yield a positive value when current is in the reference direction and a negative value when opposite to the reference.) We now verify charge and discharge directions experimentally.
    a. Move the switch to charge and note the sign of the multimeter reading. (Don't try to read its value—we are only interested in its sign.) Sign _____. Thus, the actual direction of current

    is _____.

    b. Move the switch to discharge and again note the sign. Sign

    _____. Thus, the actual direction of current is

    _____.

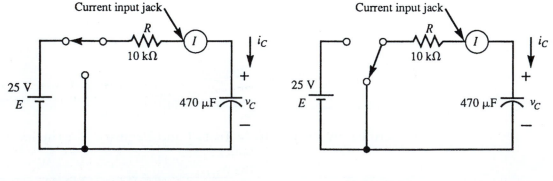

a) Charge   b) Discharge

Figure 9-4   Verifying the current direction convention

PART C: More Complex Circuits

For analysis, complex circuits can be reduced to their Thévenin equivalent. In this test, you will verify the Thévenin equivalent method.

5. a. Using Thévenin's theorem, determine the charge and discharge equivalent circuits for the circuit of Figure 9-5(a). Sketch in (b) and (c) respectively.

     b. From the Thévenin equivalents, compute the charge and discharge time constants.

$$\tau_{charge} = \underline{\hspace{3cm}}. \quad \tau_{discharge} = \underline{\hspace{3cm}}.$$

     c. For charging, compute capacitor voltage at $t = \tau_{charge}$.

$$v_C = \underline{\hspace{3cm}}.$$

     d. Assume the switch has been in the charge position long enough for the capacitor to fully charge. Now move the switch to discharge and compute the voltage after one discharge time constant. $v_C = \underline{\hspace{3cm}}.$

     e. Assemble the circuit of Figure 9-5(a). Move the switch to charge and with the stopwatch, measure how long it takes for the voltage to reach the value determined in Test 5(c).

$$\tau_{charge(measured)} = \underline{\hspace{3cm}}.$$

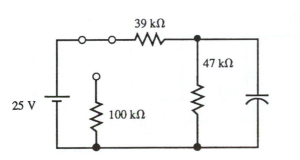

(a) Circuit. C = 470 µF　　　　　　　　*(b) Charge equivalent*　　*(c) Discharge equivalent*

Figure 9-5 Thévenin analysis

f. Allow the capacitor to fully charge, then move the switch to discharge and with the stopwatch, measure the time that it takes for the voltage to reach the value determined in Test 5(d).

$\tau_{discharge(measured)}$ = _____.

Discuss results. In particular, how well do the Thévenin equivalents represent the charging/discharging behavior of the capacitor in the real circuit?

PART D: Charging/Discharging Waveforms

> **Preliminary Note**
> Part D is set up as an instructor demonstration. However, if you have already learned how to use the oscilloscope, you may perform this part yourself.

The circuit is shown in Figure 9-6. A function generator simulates switching by applying a signal that cycles between voltage E and ground. A high rate of switching and a small time constant are used to give a waveform that can be viewed on the oscilloscope. Since a non-electrolytic capacitor is used, agreement between theory and practice will be much better than in previous tests.

6. a. Measure R and C, then assemble the circuit of Figure 9-6. Using the function generator, apply an input with 0.5 ms high and low times (i.e., f = 1 kHz) and adjust the input to 5 V as indicated in Figure 9-6. Set the oscilloscope to display only the charging voltage. Sketch as Figure 9-7.

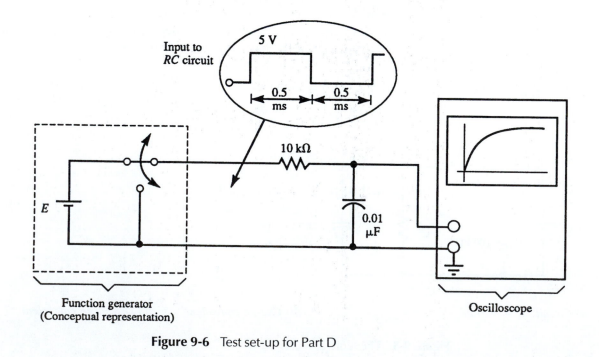

Figure 9-6 Test set-up for Part D

Figure 9-7   Waveform for Test 6.

   b. Calculate the time constant using the measured R and C.

     $\tau =$ _____

   c. From the scope screen, measure capacitor voltages at $t = \tau$, 2τ, 3τ, 4τ, and 5τ and tabulate.
   d. Using circuit analysis techniques, compute voltages at these points. How well do they agree?
   e. Display the discharge potion of the waveform and measure capitor voltages at $t = \tau$, 2τ, 3τ, 4τ, and 5τ and tabulate.
   f. Repeat Part d) for the discharge values.

COMPUTER ANALYSIS

PSPice Users
Set V1, V2, PW, and PER, TR, TF and TD to 0V, 5V, 0.5ms, 1ms, 1ns, 1ns and 0 respectively. This creates the desired waveform.

7. Using either MultiSIM or PSpice, set up the RC circuit of Figure 9-6 with a pulse source that supplies 5 V for 0.5 ms and 0 V for the next 0.5 ms as indicated in Figure 9-6. (See Section 11.8 and Figure 11-44 or 11-49(c) of the text for reference. Don't forget to use a ground on the bottom end of the source symbol.) Set the capacitor initial voltage to zero and run a transient analysis. Using the cursor, determine voltages at $t = \tau$, 2τ, 3τ, 4τ and 5τ for both the charge and discharge case. Compare the results to those obtained from the oscilloscope, Test 6.

PROBLEMS

8. Assume the circuit of Figure 9-1 has a time constant of 100 μs. If R is doubled and C is tripled, calculate the new time constant.
9. For Figure 9-1, replace the wire between switch position 2 and

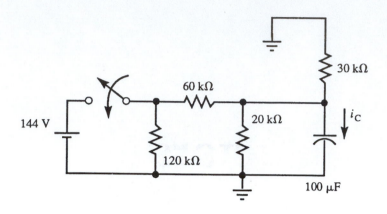

Figure 9-8   Circuit for Problem 10

common with a resistor R_2. If $R_2 = 4\,R$ and the capacitor takes 25 ms to reach full charge, how long will it take to discharge?

10. For Figure 9-8, the switch is closed at $t = 0$ s and opened 5 s later. The capacitor is initially uncharged.
   a.   Determine the capacitor current i_C at $t = 2$ s.
   b.   Determine the capacitor current i_C at $t = 7$ s.

FOR FURTHER INVESTIGATION AND DISCUSSION

Tolerances on resistors and capacitors affect the time constant of circuits and thus affect circuits that rely on RC charging and discharging for their operation. To investigate, assume the resistor of Fig. 9-6 has a tolerance of ±5% and the capacitor ±10%. Write a short discussion paper on the impact of these tolerances on rise and fall waveforms. In your discussion, determine the minimum and maximum values that τ may have, then plot the waveforms for these two cases as well as for the nominal case on the same graph. Using PSpice or MultiSIM, verify these waveforms. Discuss the ramifications of what you have learned.

NAME _____

DATE _____

CLASS _____

LAB 10

Inductors in dc Circuits

OBJECTIVES

After completing this lab, you will be able to
- determine steady state dc voltages and currents in an *RL* circuit,
- measure the inductive "kick" voltage in an *RL* circuit,
- measure the time constant of an *RL* circuit.

EQUIPMENT REQUIRED

- ☐ DMM
- ☐ Variable dc power supply
- ☐ Function generator
- ☐ Oscilloscope

COMPONENTS

- ☐ Resistors: 10-Ω, 82-Ω (two), 1-kΩ (1/4-W),
  47-Ω, 220-Ω (1/2-W)
  120-Ω (two, each 2-W)
- ☐ Inductors: One, 2.4-mH, powdered iron core, Hammond
  Part #1534 or equivalent; one approximately
  1.5-H, iron core.

Note to the Instructor

Parts C and D of this lab may be run as an instructor demo if your students have not yet had instruction on the oscilloscope.

EQUIPMENT USED

Instrument	Manufacturer/Model No.	Serial No.
DMM		
Power supply		
Function generator		
Oscilloscope		

Table 10-1

TEXT REFERENCE

Section 13.2   INDUCED VOLTAGE AND INDUCTION
Section 13.7   INDUCTANCE AND STEADY STATE DC
Section 14.1   INTRODUCTION
Section 14.2   CURRENT BUILD UP TRANSIENTS
Section 14.4   DE-ENERGIZING TRANSIENTS

DISCUSSION

Induced voltage is determined by Faraday's law. For a coil with a non-magnetic core, induced voltage is directly proportional to the rate of change of current and is given by

$$v_L = L\frac{di}{dt} \tag{10-1}$$

where L is the inductance of the coil and di/dt is the rate of change of the current in the coil. The induced voltage opposes the change in current.

For inductances in series, total inductance is the sum of individual inductances. Thus,

$$L_T = L_1 + L_2 + ... + L_N \tag{10-2}$$

For inductances in parallel, total inductance may be found from

$$\frac{1}{L_T} = \frac{1}{L_1} + \frac{1}{L_2} + ... + \frac{1}{L_N} \tag{10-3}$$

Since real inductors have coil resistance, Equation 10-3 can only be used in practice if coil resistance is negligible.

Steady State and Transient Response

Steady State dc: As Equation (10-1) shows, voltage results only when current changes. Thus if current is constant (as in steady state dc), the voltage across an inductance is zero. Consequently, *to steady state dc, an inductance looks like a short circuit.*

Current Build up Transients: Simple current build up transients in *RL* circuits may be studied using the circuit of Figure 10-1(a). When the switch is closed, current in the inductor is given by

$$i_L = \frac{E}{R_1}(1 - e^{-R_1 t/L}) \tag{10-4}$$

which has a final steady state value of E/R_1 amps as shown in (b). Voltage is given by

$$v_L = Ee^{-R_1 t/L} \tag{10-5}$$

As Figures 10-1(b) and (c) show, at the instant the switch is closed, current is zero and full source voltage appears across the inductance. This means that *an inductor, with initial current of zero amps, looks like an open circuit.*

Current Decay Transients: Consider Figure 10-2(a). Assume that the current in the inductor at the instant the switch is opened is I_0. Since current cannot change instantaneously, *a current carrying inductance looks momentarily like a current source of I_0 at the instant of switch operation.* Current then decays to zero according to

$$i_L = I_0 e^{-Rt/L} \tag{10-6}$$

where $R = R_1 + R_2$ for the circuit of Figure 10-2. Other voltages and currents can be obtained from these relationships using basic circuit principles. For example, the voltage v_2 across resistor R_2 is $-R_2 i_L$. Thus, it has the same shape as i_L but is negative as indicated in (c). (Multiplying $-R_2$ times Equation 10-6 yields $v_2 = -I_0 R_2 e^{-Rt/L}$ as indicated in Figure 10-2c.) Note that V_0 (which is equal to $-I_0 R_2$), can be many times larger than the source voltage.

The time constant of an *RL* circuit is given by

$$\tau = L/R \tag{10-7}$$

where R is the resistance through which the inductor current builds or decays; for Figure 10-2, $\tau = L/R_1$ during current build up and $\tau = L/(R_1 + R_2)$ during current decay. Steady state is reached in 5τ where the appropriate τ (charge or discharge) must be used.

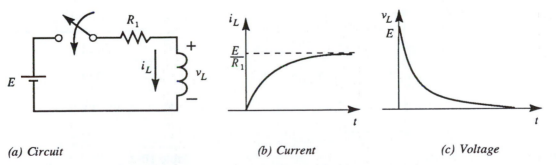

(a) Circuit   (b) Current   (c) Voltage

Figure 10-1   Current build up transients

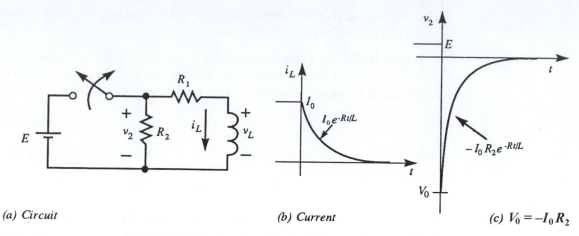

(a) Circuit   (b) Current   (c) $V_0 = -I_0 R_2$

Figure 10-2   Current decay transients. $R = R_1 + R_2$ for (a) and (b).

MEASUREMENTS

PART A: Steady State Voltages and Currents

1. Measure the resistance of each 120 Ω resistor and the resistance of the 2.4-mH inductor and record in Table 10-2.
2. a. Assemble the RL portion of the circuit of Figure 10-3. With the source disconnected, measure input resistance R_{in}.

    $R_{in(measured)}$ = _____

    b. Using circuit analysis techniques, compute R_{in}. $R_{in(computed)}$ =

    _____. Compare to the measured value.

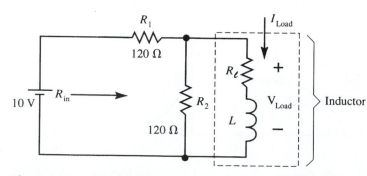

Figure 10-3   Circuit for Test 1

	Nominal	Measured
R_1	120 Ω	
R_2	120 Ω	
R_ℓ	xxxxx	

Table 10-2

c. Connect the source and measure voltage V_{Load} and I_{Load}.

$$V_{\text{Load(measured)}} = \underline{\hspace{3cm}} \qquad I_{\text{Load(measured)}} = \underline{\hspace{3cm}}$$

d. Using circuit analysis techniques, solve for V_{Load} and I_{Load}. Compare to the measured values.

PART B: More Complex Steady State Circuits

For more complex circuits, use Thévenin's theorem to reduce portions of the circuit as necessary.

3. a. Measure each resistor for the circuit of Figure 10-4 and record in Table 10-3. Use the same inductor as in Figure 10-3.
    b. Determine the Thévenin equivalent of the circuit to the left of the inductor using circuit analysis techniques using the measured resistance values from Table 10-3.

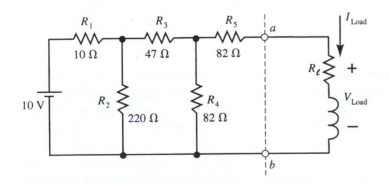

	Nominal	Measured
R_1	10 Ω	
R_2	220 Ω	
R_3	47 Ω	
R_4	82 Ω	
R_5	82 Ω	

Figure 10-4 Circuit for Test 3. Use resistor values measured in Table 10-3.

Table 10-3

c. Disconnect the inductor and measure the open circuit voltage across a-b. (This is E_{Th}.) $E_{Th(measured)}$ = _____.

d. Disconnect the source and replace it with a short circuit. Measure the resistance looking back into the circuit. (This is R_{Th}.)

$R_{Th(measured)}$ = _____

e. Compare the measured and computed values for E_{Th} and R_{Th}.

f. Remove the short, reconnect the source and inductor and measure I_{Load} and V_{Load}.

I_{Load} = _____ V_{Load} = _____

g. Using the Thévenin equivalent determined in (b), compute I_{Load} and V_{Load}. Compare to the measured values of (f).

PART C: Inductive "Kick" Voltage

Preliminary Note
Part C requires the use of an oscilloscope. If you have not yet covered the oscilloscope, this may be run as an instructor demonstration.

In this part of the lab, you will look at the *inductive kick* that results when current in an inductor is interrupted. Use the iron core inductor. (The inductor and resistor values used here are not critical. However, to see the results easily on the scope, you need a time constant of at least a few milliseconds. First determine appropriate resistor values.)

4. a. Measure the dc resistance of the inductor. R_ℓ = _____.
Consider Figure 10-5(a). Define $R_1' = R_1 + R_\ell$. Select a value for R_1 such that current E/R_1' is easily handled by your power supply. Now select an R_2 that is about five or six times larger than R_1' and such that the discharge time constant $L/(R_1 + R_2 + R_\ell)$ is a few milliseconds or more. (If you know L, you can compute τ; if you have no way of measuring L, experiment until you get a waveform you can see.)

b. Assemble the circuit with these values and connect the scope probe across resistor R_2 to view the inductive kick voltage v_2.

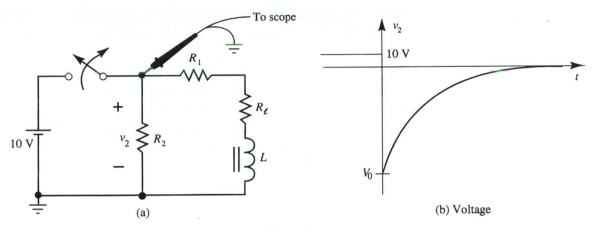

Figure 10-5   Circuit for Test 4

(The waveform will look like that shown in Figure 10-5(b). Voltage V_0 depends on the ratio of R_2 to R_1'. If $R_2 = 5\, R_1'$ and $E = 10$ V, then $V_0 = -50$ volts.) The waveform is a single shot event that is generated only when you open the switch. Set the scope triggering appropriately—e.g., *Norm* or *Single Sweep*, negative slope. With careful adjustment of triggering and the time base and with repeated operation of the switch, you should be able to observe the voltage spike of (b).

c. From the trace, measure the value of V_0. $V_0 = $ _____

d. Using the methods of Section 14.4 of the textbook, compute V_0 and compare to the measured value of (c). How do they compare?

PART D: Measuring the Time Constant

This part may be run as a demo if necessary.

5. a. Measure the resistance of the 1-kΩ resistor and the inductance of the 2.4-mH inductor. (Call the 1-kΩ resistor R_S.) If you do not

have any way to measure inductance, use the nominal value marked on the inductor. Use R_ℓ from Table 10-2.

R_ℓ = _____ L = _____ R_S = _____

b. Assemble the circuit of Figure 10-6. The time constant for the circuit is $\tau = L/R_T$ where R_T is the sum of R_S, R_ℓ and the output resistance R_{out} of your function generator. (This may be determined from the front panel of the generator. A typical output resistance of a function generator is 50 Ω.) Thus,

τ = _____ .

c. Set the function generator to the square wave mode at 40 kHz. (This provides ample time for current to build up and decay fully.) Adjust the signal generator and scope so that the amplitude of the v_S waveform is 5 grid lines (i.e., 5V). The waveform should look like that shown in Figure 10-6.
d. Adjust the time base and triggering to get only the build up waveform on the screen. Sketch this waveform as Figure 10-7.
e. Measure the time that it takes for the waveform to reach 63.2% of its final value. This is the measured time constant.

$\tau_{measured}$ = _____

Compare this value to the value computed in (b).
f. Change triggering to get the decay waveform on the screen. Measure the time constant here and compare to that of (d).

$\tau_{measured}$ = _____

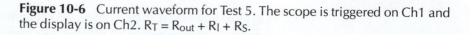

Figure 10-6 Current waveform for Test 5. The scope is triggered on Ch1 and the display is on Ch2. $R_T = R_{out} + R_l + R_S$.

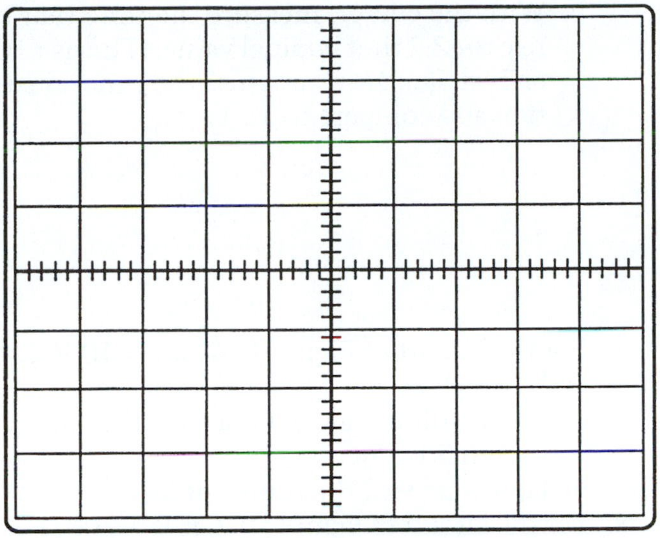

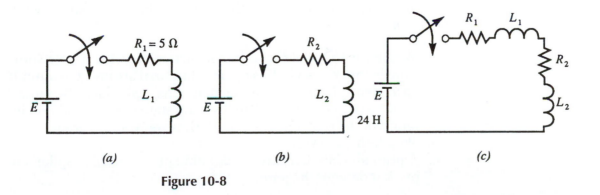

Figure 10-7   Waveform for Test 5

COMPUTER ANALYSIS

6. Using either MultiSIM or PSpice, set up the circuit of Figure 10-6. To create a waveform equivalent to that used in Test 5, use a 40 kHz square wave with equal high and low times. If you are using PSpice, use pulse source VPULSE, whereas, if you are using Multi-SIM, use the clock. (For reference, see Section 11.8 of the text.) To determine the amplitude for this pulse, note that in Test 5(c), the peak voltage across R_S is 5 V. This means that the peak current must be 5 mA. Use this to compute E. For example, if R_{out} = 50 Ω, coil resistance is 12 Ω and R_S = 1 kΩ, then you need a pulse amplitude of $E = (5 \text{ mA})(50 \text{ Ω} + 12 \text{ Ω} + 1 \text{ kΩ}) = 5.31$ V.

a. Run a transient analysis with the initial inductor current set to zero.

Figure 10-8

b. With the cursor, measure the time that it takes for current to rise to 63.2% of its final value. (This is τ.) Compare to the result of Test 5(e). Now measure the time constant on the decay portion and compare to Test 5(f).

PROBLEMS

7. For the circuit of Figure 10-2, if $E = 10$ V, $R_1 = 200\ \Omega$, $R_2 = 1200\ \Omega$ and $L = 5$ H,
   a. What voltage appears across the switch at the instant the switch is opened?
   b. How long will the transient last?
8. For the circuit of Figure 10-2, if $R_1 = 200\ \Omega$, $R_2 = 100\ \Omega$ and $L = 5$ H,
   a. What voltage appears across R_2 at the instant the switch is opened?
   b. How long will the transient last?
9. The circuit of Figure 10-8(a) takes 4 s to reach steady state, while that of (b) takes 12 s. The circuits of (a) and (b) are combined as in (c). How long will it take for the circuit of (c) to reach steady state?

FOR FURTHER INVESTIGATION AND DISCUSSION

Remember
Do not perform a full analytic transient analysis to determine the required information

When you disturb an RL or an RC circuit you trigger a transient, creating voltages and currents that may greatly exceed the circuit's normal steady state values. Many times, however, it is only necessary to determine the general nature of the resulting waveforms and find values at key points—i.e., you do not need a full transient solution. To help develop this idea, consider Figure 10-4. First, replace the 10 V source with a 250-mA dc current source, then install a switch in the leg containing R_4. Assume the coil resistance is negligible. Initially the circuit is in steady state with the switch closed. You then open the switch. Prepare a written analysis of what happens using the following for guidelines.

a. Sketch the shape of the inductor current waveform from some time before the switch is operated to final steady state circuit operation. Using a series of diagrams, calculate the initial steady state current, the final steady state current and the duration of the transient, and mark these on your diagram.
b. Sketch the coil voltage waveform.
c. (Optional) Calculate the peak value of the voltage spike and mark it on your diagram.

The Oscilloscope (Part 1) Familiarization and Basic Measurements

OBJECTIVES

After completing this lab, you will be able to
- describe the operation and usc of an oscilloscope,
- connect an oscilloscope to a circuit under test and select basic control settings,
- measure dc voltage,
- use an oscilloscope to observe time varying waveforms.

EQUIPMENT REQUIRED

☐ Oscilloscope
☐ Variable dc power supply
☐ Function generator
☐ DMM

Preliminary Note

Basic features of the oscilloscope are covered in *A Guide to Lab Equipment and Laboratory Measurements* at the beginning of this manual. You may wish to review this material before doing the lab.

EQUIPMENT USED

Instrument	Manufacturer/Model No.	Serial No.
DMM		
Power supply		
Function generator		
Oscilloscope		

Table 11-1

DISCUSSION

The oscilloscope is the key test and measurement instrument used for studying time varying waveforms. Its main feature is that it displays waveforms on a screen; with an oscilloscope, you can view and study waveforms, measure ac and dc voltages, frequency, period, phase displacement and so on. However, the oscilloscope is a fairly complex instrument and we therefore learn about it in stages. In this lab, we concentrate on operational procedures, front panel controls and a few basic measurements; in Labs 12 and 13, we look at more advanced measurement techniques. Later labs add more detail.

Connecting to the Circuit Under Test

The oscilloscope is connected to the circuit under test by means of a probe (or set of probes) as illustrated in Figure 11-1. The probe includes a measurement tip and a ground clip and connects to the oscilloscope via a flexible, shielded cable which is grounded at the oscilloscope. This ground serves as the reference point with respect to which all signals are measured. The shield helps guard against electrical noise pick up.

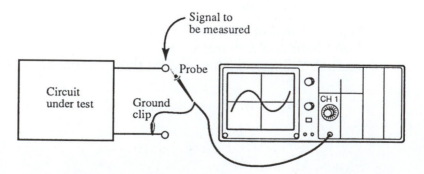

Figure 11-1 The basic oscilloscope measurement circuit

Probes may be "× 1" or "× 10". A "× 10" probe contains a 10:1 voltage divider which attenuates the signal by a factor of 10; thus, when you use a "× 10" probe, you have to multiply the scope readings by a factor of 10 to get the correct input (unless your scope automatically changes scales for you as some models do).

Front Panel Controls

Front panel controls permit you to control the operation of the oscilloscope. They may be grouped functionally as in Table 11-2.

Main Oscilloscope Controls According to Function			
Display	**Vertical**	**Horizontal**	**Triggering**
Intensity	Coupling (ac-gnd-dc)	Sec/Div	Coupling
Focus	Volts/Div	X-position	Source
Beam Finder	Y-position		Level
	Channel Select		Slope
			Mode

Table 11-2

A few of these are summarized below. Others will be introduced in later labs.

Coupling (ac-gnd-dc): Permits selection of coupling. When set to *dc*, the entire signal (ac plus any dc present) is displayed; when set to *ac*, dc signals are blocked by a capacitor and only ac is displayed; when set to *gnd*, the input is grounded. This permits establishing a *0-V base-line* (reference) on the screen.

VOLTS/DIV: This is the scope's vertical sensitivity control. It is a calibrated control that establishes how many volts each major vertical scale division represents. For example, when set for 10-V/DIV, each grid line represents 10 volts. Each channel has its own independent VOLTS/DIV control. A fine adjust control is also provided, but it is not calibrated. Each scale division is usually 1 cm.

Y-Position: This is the vertical position control. Each channel has its own control. It moves the trace up or down for easier observation. It is not calibrated.

Channel Select: Permits displaying Ch1, Ch2, both, or their sum or difference.

SEC/DIV: This is a calibrated control that selects how many seconds each major horizontal division represents. (It is calibrated in s, ms, and μs.) One control handles all channels. (A non-calibrated fine adjust is also provided.)

X-Position: Positions the trace horizontally. One control handles all channels.

Trigger Source: Selects the trigger source, e.g., Ch1, Ch2, an external trigger, or the ac line.

Trigger Level: Permits you to adjust the point on the trigger source waveform where you want triggering to start.

Trigger Slope: Selects whether the scope is to trigger on the positive or negative slope of the trigger source waveform.

Trigger Mode: Modes include *auto, normal* and *single sweep.* In the auto mode, the sweep always occurs, even with no trigger present; in the normal mode, a trigger must be present; in the single sweep mode, a trigger is required but only one sweep results. (Other modes may be provided but we will not consider them here.)

MEASUREMENTS

PART A: General Familiarization

If you have trouble getting a trace on the screen, check the intensity control; if it is set too low, the trace may be very faint or not visible. *(Caution: Never leave a bright spot on the screen.)* If adjusting the intensity does not locate the trace, proceed as follows: select Ch1, set triggering to auto, set SEC/DIV to mid range, press and hold the *beam finder* control, then adjust the vertical control to locate the trace.

Tests 1 to 3 are performed with no input applied to the oscilloscope.

1. Rotate the *focus* and *intensity* controls and note their effect. Adjust until you get a sharply focused trace at a comfortable viewing level.
2. Adjust the *vertical position* control and note its effect. Center the trace vertically on the center.
3. Adjust the *horizontal position* control and note its effect. Center the trace horizontally on the screen.
4. Connect a probe to Ch1 and set the channel selector to Ch1. Touch the probe tip to the *calibration test point* on the front panel. (Do not connect the ground clip.) Adjust the *VOLTS/DIV* control, the *SEC/DIV* control and the *trigger controls* until you get the calibration waveform on the screen. (It should be a square wave.)

PART B: Measuring dc Voltage with the Oscilloscope

5. a. Set the channel selector to Ch1 and use a ×1 probe. (Ensure the *VOLTS/DIV* switch for Ch1 is on *CAL.*) Set the trigger to *auto.* Move the *ac-gnd-dc* switch to *gnd* and center the trace. Return the coupling switch to *dc.* Voltage can be

> **Caution**
> The ground points on oscilloscopes, power supplies, and other equipment are generally tied to the electrical power system
>
> (continues next page)

determined from the screen using the relationship $V = (deflection) \times (VOLTS/DIV\ setting)$.

b. Connect the probe as in Figure 11-2 and set *VOLTS/DIV* to 1 V. With the voltmeter, set the power supply to 2 V and note the deflection on the screen. From this deflection, compute the measured voltage. (It should equal the applied voltage.) Record in Table 11-3.

c. Now change *VOLTS/DIV* to 2 V, set $E = 4$ V and note the position of the trace. Enter data in Table 11-3 and compute *V*. Repeat for $E = 15$ V at 5 *VOLTS/DIV*.

d. Replace the probe with an ×10 probe. Using the oscilloscope, set the supply successively to 10 V, 15 V and 22.5 V. Record data, including the *VOLTS/DIV* settings that you choose. Compare to the meter reading.

ground through the U-ground pin on the electrical power outlet (plug-in). Since this connection ties all ground points together, you must be careful when connecting ground clips, as it is easy to inadvertently short out a component or even accidentally ground an output. While these grounds are required for safety reasons, they make poor signal paths. Therefore, be sure to use the ground clip supplied with the scope probe when making measurements.

Probe	Input Voltage	VOLTS/DIV Setting	Deflection	Voltage from Oscilloscope
× 1	2 V	1-V		
× 1	4 V	2-V		
× 1	15 V	5-V		
× 10				10 V
× 10				15 V
× 10				22.5 V

Table 11-3

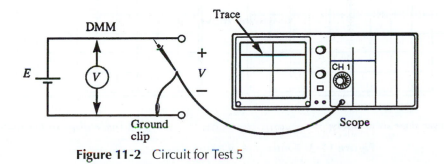

Figure 11-2   Circuit for Test 5

6. Move the input coupling switch to *ac*. Set the supply to the voltages of Table 11-3. What happens? Why?

    _____

7. Return coupling to *dc*. Now set the trigger to *normal*. Note that the trace disappears. (This is because there is no trigger point in dc to start the sweep.) Return the trigger to *auto*.
8. Make sure your power supply output is floating. Reverse the probe connections to the power supply. Adjust the power supply voltage up and down. Note the deflection on the screen. Describe what happened.

    _____

PART C Observing Waveforms

Replace the power supply with a function generator. Be sure to connect the ground of the scope to the ground of the generator. Set input coupling to *gnd* and center the trace. Change coupling to *ac*.

9. a. Set the function generator to a 2-kHz sine wave. On the oscilloscope, set the *VOLTS/DIV* switch to 1 V, the trigger to positive slope, and the time base to 0.1 ms/div. Adjust the output voltage of the generator until you get a nicely sized sine wave on the screen, then trim the frequency and adjust the horizontal position and trigger level controls to get one cycle of the waveform to fit between 5 horizontal grid lines. Sketch the waveform as Figure 11-3(a).
    b. Change the trigger slope to negative and repeat (a). Sketch the waveform as Figure 11-3(b).

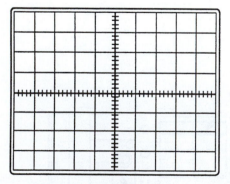

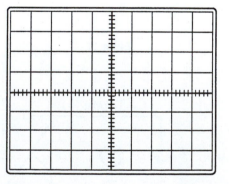

(a) Trigger slope set to positive   *(b) Trigger slope set to negative*

Figure 11-3   One cycle of a sine wave

c. Return the trigger slope to positive, then set input coupling to *dc*. (Make sure the dc offset on your function generator is set to zero.) What happens to the waveforms? (Compare to those of Figure 11-2.)

_____

d. Return the coupling to *ac*. Set *f* = 500 Hz and change the time base to get 4 cycles on the screen (actually a bit more than four). Sketch here. Be sure to note the time base setting.

10. Repeat Steps 9(a) and (b) for a square wave and for a triangular wave. Sketch waveforms as Figures 11-4 and 11-5.

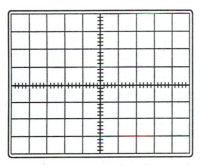
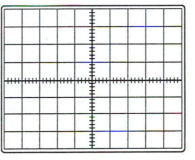

(a) Trigger slope set to positive   (b) Trigger slope set to negative

Figure 11-4   One cycle of a square wave

(a) Trigger slope set to positive   (b) Trigger slope set to negative

Figure 11-5   One cycle of a triangular wave

11. Apply a 20 kHz sine wave and with the time base on *CAL*, display a few cycles on the screen. Now take the time base control off *calibrate* and vary it. Discuss what happens _____

_____

_____

_____

PROBLEMS

12. A waveform is faint. What control should be adjusted? _____

13. A waveform is fuzzy. What control should be adjusted? _____

14. With a ×1 probe and *VOLTS/DIV* (vertical sensitivity) set at 5 V/div, a dc voltage moves the trace up 3.4 grid lines. What is the input voltage? _____

15. With a ×1 probe and the vertical sensitivity set at 10 V/div, a dc voltage moves the trace down by 1.2 grid lines. What is the input voltage? _____ .

16. With a ×10 probe and the vertical sensitivity control set at 5 V/div, a dc voltage moves the trace up 2.5 grid lines. What is the input voltage? (There are two possible answers, depending on your oscilloscope—see earlier note. Answer in terms of the scope you used in this lab.)

17. Using a ×1 probe and with the vertical sensitivity set to 2 V/div, a dc voltage moves the trace up by 2 and a half grid lines. The *VOLTS/DIV* fine adjust is used to bring the trace up to 3 grid lines high. What is the value of the input voltage? Why?

Basic ac Measurements: Period, Frequency, and Voltage (The Oscilloscope—Part 2)

OBJECTIVES

After completing this lab, you will be able to use an oscilloscope to
- measure period and frequency of an ac waveform,
- measure amplitude and peak-to-peak voltage,
- measure instantaneous voltage,
- determine the equation for a sinusoidal voltage from the oscilloscope readings.

EQUIPMENT REQUIRED

☐ Oscilloscope
☐ Signal or function generator

EQUIPMENT USED

Instrument	Manufacturer/Model No.	Serial No.
Oscilloscope		
Signal or function generator		

Table 12-1

TEXT REFERENCES

Section 15.4 FREQUENCY, PERIOD, AMPLITUDE, AND PEAK VALUE

Section 15.5 ANGULAR AND GRAPHIC RELATIONSHIPS FOR SINE WAVES

Section 15.6 VOLTAGES AND CURRENTS AS FUNCTIONS OF TIME

> **Note**
> Higher end oscilloscopes can do some of the tasks described here automatically, for example, they can determine the frequency (and other waveform parameters) and display them digitally on the screen. However, we will limit our discussion to the basic functions that are common to all oscilloscopes.

Measuring Period and Frequency: The period of a waveform is the length of one cycle. Since the horizontal scale of an oscilloscope is calibrated in seconds, you can measure the period T directly on the screen, then determine frequency from the relationship $f = 1/T$. For example, if the time base is set to 20 µs per division and one cycle is 4 divisions, then $T = 4(20 \, \mu s) = 80 \, \mu s$ and $f = 1/80 \, \mu s = 12.5 \, kHz$.

Measuring Voltage: An oscilloscope displays the instantaneous value of its input voltage. Thus, an oscilloscope may be used to measure peak voltage, peak-to-peak voltage, and indeed, the voltage at any point on a waveform. This voltage is measured in the same manner as dc voltage—you determine the deflection of the trace at that point, then multiply by the vertical sensitivity setting.

Equations for Sinusoidal Voltage from Oscilloscope Readings: Mathematically, the voltage at any point on a sine wave can be found from the equation

$$v = V_{m} \sin \alpha \qquad (12\text{-}1)$$

where α is the angular position on the cycle as indicated in Figure 12-1. If you know V_{m}, you can determine the voltage at any position by direct substitution into Equation 12-1. (Since one cycle represents 360°, one half cycle represents 180°, one quarter cycle represents 90°, and so on. The angular position at any other point can be determined by direct proportion.)

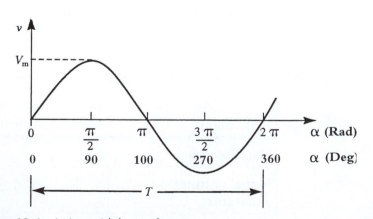

Figure 12-1 A sinusoidal waveform

Practical Note
The frequency scale of many signal generators is not very accurate and you may have to trim the frequency to get the desired period on the oscilloscope.

Voltage as a Function of Time: Equation 12-1 may be rewritten as a function of time as

$$v(t) = V_m \sin\omega t \qquad (12\text{-}2)$$

where $\omega = 2\pi f = 2\pi/T$ and t is time measured in seconds. By measuring V_m and T on the screen, you can use Equation 12-2 to establish the analytic expression for any measured sinusoidal voltage.

Control Settings: Always select *VOLTS/DIV* and *SEC/DIV* settings to yield best results. For example, if you set the amplitude and time scales to spread a waveform over the screen, you can get more accurate measurements than if you compress the waveform into a small space. (Most settings here and in future labs will be left for you to select.) In addition, make sure that you have established an appropriate zero volt base line for each channel and that the *VOLTS/DIV* and *SEC/DIV* switches are set to their *CAL* (calibrate) positions.

MEASUREMENTS

PART A: Measuring Period and Frequency

1. a. Connect the oscilloscope to the signal generator. Set the oscilloscope time base to 0.1 ms/div, coupling to *ac,* and the trigger slope to *positive.* Adjust the signal generator to obtain a sine

wave that is 5 divisions in length. Thus, $T =$ _____ and

$f =$ _____. Compare f to the frequency set on the genera-

tor dial.

_____

   b. Repeat for a time base setting of 20 μs per division and two cy-

cles in 8 divisions. $T =$ _____ and $f =$ _____. Com-

pare to the frequency set on the signal generator dial.

_____

PART B: Amplitude and Peak-to-Peak for a Sine Wave

2. For best results, set peak-to-peak amplitude rather than zero to peak. Use the vertical control to position the waveform between grid lines. For this test, set the vertical sensitivity to 0.5 V/div and choose a frequency of 1 kHz.

a. Adjust the signal generator to yield a display of 8 grid lines peak-to-peak, centered vertically. What is peak-to-peak voltage? $V_{peak-to-peak}$ = _____.

b. What is V_m? V_m = _____.

c. Sketch the waveform below with V_m and peak-to-peak voltages carefully labeled.

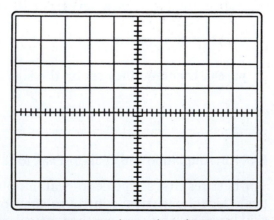

PART C: Instantaneous Value of a Sine Wave

3. a. Set the vertical sensitivity control to 1 V/div, the time base to 0.1 ms/div and obtain a waveform on the screen with a cycle length of exactly 8 divisions. Adjust the signal generator for a peak-to-peak display of 8 grid lines. Sketch the waveform as Figure 12-2. Label the vertical and horizontal axes in volts and ms.

b. From the screen, measure the voltage at 0.1 ms intervals and record in Table 12-2.

Figure 12-2   Measured waveform for Test 3

c. How many degrees does each 0.1 ms division represent?

_____.

Record the value of α for each value of t shown in Table 12-2.

d. Using Equation 12-1, verify each entry in the table. (Show a few sample calculations.)

t (ms)	α (deg)	Voltage	
		Measured	Computed
0			
0.1			
0.2			
0.3			
0.4			
0.5			
0.6			
0.7			
0.8			

Table 12-2

PART D: The Equation for Sinusoidal Voltage

4. a. For the waveform of Figure 12-2, determine ω.

$\omega =$ _____

b. Using the measured values of ω and V_m, write the equation for the voltage in the form of Equation 12-2.

$v(t) =$

c. Using this equation, compute v at $t = 50\,\mu s$ and $t = 150\,\mu s$. Show details below.

PROBLEMS

5. With the time base set to *0.5 µs/DIV*, four cycles of a waveform occupies 10 divisions. What is the period and the frequency of the waveform?

   Period _____ Frequency _____

6. With the *VOLTS/DIV* set to 2 and a ×1 probe, a waveform has an amplitude of 2 1/2 grid lines. The *VOLTS/DIV* fine adjust is used to bring the amplitude up to 3 grid lines high. What is the amplitude of the input voltage?

   Amplitude _____

7. Given $v(t) = 100 \sin 377t$

   a. What is the value of V_m? $V_m =$ _____

   b. What are the frequency and period?

      $f =$ _____ $T =$ _____

   c. Compute the voltage at $t = 20$ ms. Sketch the waveform and show where $t = 20$ ms occurs on the waveform.

8. Determine the equation $v(t)$ for the voltage depicted in Figure 12-3.

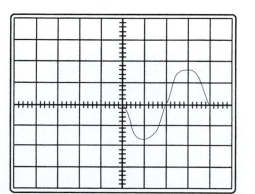

Vertical = 5V/div
Time base = 0.1 µs/div

Figure 12-3   Assume the waveform starts at time $t = 0$s

AC Voltage and Current (The Oscilloscope- Part 3, Additional Measurement Techniques)

OBJECTIVES

After completing this lab, you will be able to use an oscilloscope to
* measure rms values for sinusoidal voltage,
* measure superimposed ac and dc voltages,
* measure ac current using a sensing resistor,
* display two waveforms simultaneously on a dual channel oscilloscope,
* measure phase displacement with a dual channel oscilloscope,
* measure voltage using differential measurement techniques.

EQUIPMENT REQUIRED

☐ Dual channel oscilloscope
☐ Signal or function generator
☐ DMM and VOM

COMPONENTS

☐ Resistors: 82-Ω, 100-Ω, 150-Ω, 1-kΩ, 2-kΩ, 3.3-kΩ, 5.6-kΩ, 6.8-kΩ, 10-kΩ (all 1/4-W)
☐ Capacitor: 0.01-μF
☐ Battery: 1.5-V

EQUIPMENT USED

Instrument	Manufacturer/Model	Serial No.
DMM		
VOM		
Dual Channel oscilloscope		
Signal of function generator		

Table 13-1

TEXT REFERENCE

Section 15.6   VOLTAGES AND CURRENTS AS FUNCTIONS OF TIME

Section 15.9   EFFECTIVE VALUES

DISCUSSION

Measuring the rms Voltage of a Sine Wave. An oscilloscope may be used to determine the rms value of a sinusoidal voltage since, for a sine wave, $V_{rms} = 0.707\ V_m$ and V_m can be measured directly on the screen. One of the reasons you might do this is convenience—if you have a signal already displayed on the screen, there is no need to connect a voltmeter. However, a more fundamental reason has to do with frequency—most meters have very limited frequency ranges. For example, many DMMs can measure to only a few kHz (although some can measure to much higher frequencies), while analog VOMs typically can measure up to about 100 kHz. (Check your meter's specs.) On the other hand, even inexpensive oscilloscopes can measure to the tens of MHz, while top of the line units can measure to hundreds of MHz, or even to the GHz range.

Measuring Current. Oscilloscopes can also be used to measure current, although not directly. There are however, two indirect ways to measure current. The first is to insert a known resistor (sometimes called a *sensing resistor*) into the circuit, measure the voltage across it using an oscilloscope, then use Ohm's law to compute current. (This method is inexpensive and widely used. The other method uses a *current probe*, but few introductory courses have access to equipment of this type so we won't consider it.)

The current sensing resistor approach is based on Ohm's law. For a purely resistive circuit, $v = Ri$. Thus, if voltage is sinusoidal, current will be sinusoidal also and vice versa—that is, v and i are in phase. Therefore, if voltage v is

$$v = V_m \sin \omega t$$

then $\qquad\qquad\qquad\qquad i = I_m \sin \omega t$

Current can thus be determined by measuring V_m and computing I_m from $I_m = V_m/R$. If the rms value of current is needed, it may be determined from the equation $I_{rms} = 0.707\, I_m$.

Dual-Channel Measurements. With a dual channel oscilloscope, you can display two waveforms simultaneously. This permits you to determine phase relationships between signals, compare wave shapes, and so on.

Differential Voltage Measurements. Sometimes you need to measure the voltage across a component where the normal technique of placing the probe tip at one end and the ground clip at the other end shorts out part of the circuit. For problems such as this, *differential measurement* can be used. In PART E of this lab, you learn how make such measurements.

MEASUREMENTS

PART A: RMS Values and the Frequency Response of ac Meters

We begin with a look at the frequency response of various instruments. Here, we compare the ability of the oscilloscope, DMM, and VOM to measure voltage at different frequencies.

1. a. Assemble the circuit of Figure 13-1. Adjust the signal generator to a 100 Hz sine wave with 12 V peak-to-peak (i.e., $V_m = 6$ V. Thus, $V_{rms} = 0.707 \times 6 = 4.24$ V. This is recorded in Table 13-2 as *Actual rms.*) Now measure and record the rms voltage using the DMM and the VOM.

    b. Repeat step (a) at the other frequencies indicated in Table 13-2.

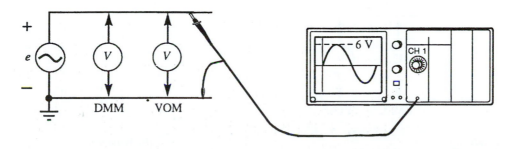

Figure 13-1 Circuit for Test 1

Frequency	Scope Reading	Acutal rms	DMM reading	VOM reading
100 Hz	6 V	4.24 V		
1000 Hz	6 V	4.24 V		
10000 Hz	6 V	4.24 V		
100 kHz	6 V	4.24 V		
1 MHz	6 V	4.24 V		

Table 13-2

c. What conclusion do you draw from the data of Table 13-2?

_____

_____

PART B: Superimposed ac and dc

2. Add the 1.5-V battery to the circuit as in Figure 13-2(a). Set the signal generator to a 100-Hz sine wave. Select *ac* coupling (to temporarily block the dc component while you set the ac component) and adjust the output of the generator to $V_m = 2$ V (i.e., 4 V_{p-p}). Return coupling to *dc*.
   a. You now have superimposed ac and dc voltages. Sketch as Figure 13-2(b).
   b. Compute the true rms value for this waveform from $V = \sqrt{V_{dc}^2 + V_{ac}^2}$ where V_{dc} is the dc component of the waveform and V_{ac} is the rms value of its ac component.

   $V =$ _____.

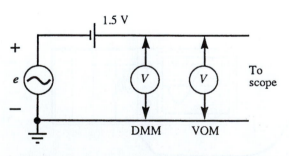

(a) Circuit   *(b) Waveform*

Figure 13-2   Circuit for Test 2

c. Measure the voltage using the meters.

DMM reading _____ VOM reading _____

d. Discuss the results of (b) and (c). In particular, why do the two meters not yield the rms value of the waveform? What type of meter is needed?

_____

_____

_____

PART C: Dual Channel Measurements

We now learn how to display two waveforms simultaneously.

3. Assemble the circuit of Figure 13-3. Set the channel select switch to *alt* and trigger mode to *auto*. (*Alt* lets you display both Ch1 and Ch2 simultaneously.)
    a. Establish the 0-V baseline for Ch1 by moving its *ac-gnd-dc* switch to *gnd* and adjust its vertical position control until the trace is centered. Repeat for Ch2. (The traces should now be superimposed.) Return both Ch1 and Ch2 to the *ac* position and set triggering to Ch1, positive slope. Set *SEC/DIV* to 50 µs/div.
    b. Set the signal generator to a 2.5-kHz sine wave with V_m = 3 V (i.e., 6 V p-p). (You should now have two sine waves on the screen.) Select an appropriate setting for Ch2 and trim the vertical position controls if necessary to ensure that both waveforms are centered about the horizontal axis.
    c. Set the time base to 20 µs/div, and using the time base variable control if necessary, spread one half cycle of the reference waveform (Ch1) over the entire screen. (This results in each

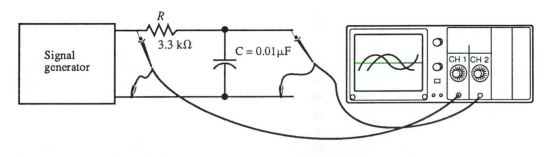

Figure 13-3   Circuit for Test 3. Use R = 3.3-kΩ and C = 0.01 µF.

major scale division representing 18°.) Now measure the displacement between the waveforms at their zero-crossover points and convert this displacement to degrees.

d. As you will learn later, the theoretical displacement is $\theta = \tan^{-1}(1/\omega RC) - 90°$. Compute θ and compare to the result of (c).

e. Measure the magnitudes of each waveform. Taking the voltage of Ch1 as reference, sketch the waveforms in this space.

f. Write the equations for the two voltages, using the values measured above.

$\mathbf{v}_1(t) =$

$v_2(t) =$

g. Set f = 10 kHz and again measure displacement. Using the equation of Test 3(d), compute displacement and compare it to the measured value.

h. Repeat Test (g) at a frequency of 18 kHz.

PART D: Current Measurement with an Oscilloscope

Consider Figure 13-4(a). The load current I_L is given by $I_L = E/R_L$ where E and I_L are the rms values of the source voltage and load current respectively. To measure this current using an oscilloscope, you can add a sensing resistor R_S as in Figure 13-4(b), then measure the voltage across it, convert to rms, then compute current as $I_L' = V_S/R_S$. (As long as R_S is small compared to R_L, the accuracy will be good.)

4. a. Accurately measure the 100-Ω sensing resistor and R_L. Set up the circuit of Figure 13-4(b). (Note the placement of the sensing resistor. With the resistor placed as shown, you can connect both ground clips of the oscilloscope directly to the ground of the signal generator without fear of ground problems.) Use a 100-Hz sinusoidal source. Using Ch1, set E_m to 3 V (i.e., 6 V p-p).

    b. Measure the voltage across the sensing resistor using Ch2, then convert to rms.

    $V_S =$ _____ (rms)

    c. Using Ohm's law, determine the rms value of the measured load current

    $I_L' = V_S/R_S =$ _____ (mA rms)

    Verify this current by comparing it to that measured by the meter.

    d. The load current that you are trying to measure is $I_L = E/R_L$, where E is the rms value of the source voltage. Using the mea-

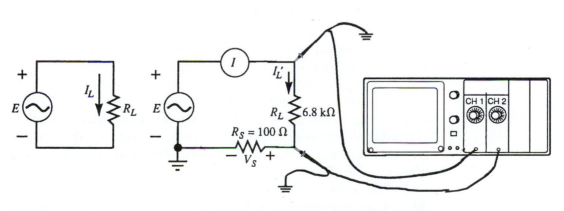

(a) Circuit   *(b) Measuring current. Here, $I_L' \approx I_L$.*

Figure 13-4   Circuit for Test 4. Use $R_S = 100$ Ω.

sured value of R_L, compute I_L using this formula and compare to the current determined by the sensing resistor approach in (c).

5. Replace R_L of Figure 13-4 with the network of Figure 13-5 and repeat steps 4(b) to 4(d).

$V_S =$ _____ (rms voltage across R_S); $I_L' = V_S/R_S =$ _____

$I_L =$ _____ (Determined by circuit analysis)

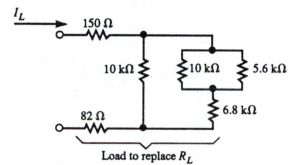

Load to replace R_L

Figure 13-5   Circuit for Test 5. Replace R_L with this network.

How do the results compare?

PART E: Differential Measurements

Consider Figure 13-6. Suppose you want to measure the voltage across resistor R_1 using an oscilloscope. As a first thought, you might try the connection shown. With this connection, however, the ground lead shorts out out R_2. Alternatively, you might try reversing the probe tip and ground clip connections. However, this shorts the source output to ground. Thus, neither approach is usable. (Of course, you might consider isolating the scope or the source from ground, but this is not desirable either, as you lose the earth safety ground.) A better approach is to use differential measurement as illustrated in Figure 13-7.

In Figure 13-7, Ch1 measures the voltage from point a to ground while Ch2 measures the voltage from point b to ground. The oscilloscope has a mode (its *differential mode*) that permits you to display Ch1 minus Ch2. This yields a display of v_{ab}, the voltage between points a and b.

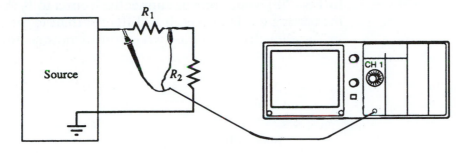

Figure 13-6 An incorrect way to view the voltage across R₁

6. a. Assemble the circuit of Figure 13-7. Set the signal generator to a 1 kHz sine wave.
   b. Set *VOLTS/DIV* to the same value for both channels and ensure that they are in the *CAL* position. Center the traces as in Step 3(a). Return the *ac-gnd-dc* switch to *ac* for both channels.
   c. Select Ch1 and set the input voltage to 3 V (i.e., 6 V p-p).
   d. Select the *difference mode* for your oscilloscope. (Details differ for different scopes. For example, some scopes require that you invert Ch2 and add it to Ch1. Check your scope's manual for details or ask your instructor.)
   e. You should now have a display of v_{ab} on your screen. Since $R_1 = 2/3\, R_T$, v_{ab} should be 2/3 of the source voltage. Verify that it is.
7. We will now look at what happens if you had tried to use the circuit of Figure 13-6 to measure the voltage across R_1.
   a. Consider again the circuit of Figure 13-7. Disconnect Probe 2 (i.e., Ch2) and set the channel select back to Ch1. You should see 3 V (i.e., 6 V p-p) on the screen since Ch1 is still measuring full source voltage.
   b. Move the ground clip of Probe 1 to point b and note the scope display.

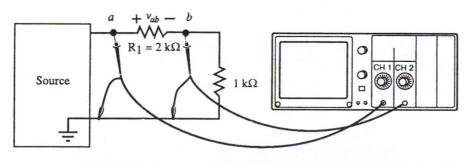

Figure 13-7 The correct way to view the voltage using differential measurement

c. In Test 7(b), your probe is connected from a to b; thus, you are measuring v_{ab}. However, it is different than v_{ab} you measured in Test 6(e). Briefly discuss what you are now seeing.

PROBLEMS

8. Given: $v_1(t) = 100 \sin \omega t$ and $v_2(t) = 80 \sin(\omega t+30°)$ displayed on a scope screen with v_1 as reference. If $f = 100$ Hz, sketch the oscilloscope trace in the space below as Figure 13-8(a).
9. Repeat Question 8 (as Figure 13-8(b)) if v_2 is the reference waveform.

Figure 13-8   (a) Waveform for Question 8   (b) Waveform for Question 9

Capacitive Reactance

OBJECTIVES

After completing this lab, you will be able to
- measure phase difference between voltage and current in a capacitance,
- measure capacitive reactance and verify theoretically,
- determine the effect of frequency on capacitive reactance.

EQUIPMENT REQUIRED

- ☐ Dual channel oscilloscope
- ☐ Signal or function generator
- ☐ DMM (two)

COMPONENTS

- ☐ Resistors: 10-Ω, 220-Ω, 1/4-W
- ☐ Capacitors: 1.0-μF (two), non-electrolytic

EQUIPMENT USED

Instrument	Manufacturer/Model No.	Serial No.
DMM #1		
DMM #2		
Dual channel oscilloscope		
Signal or function generator		

Table 14-1

TEXT REFERENCE

Section 16.6   CAPACITANCE AND SINUSOIDAL AC

DISCUSSION

Capacitance and Sinusoidal ac: When an ideal capacitance is connected to a sinusoidal voltage source, current leads by 90° as illustrated in Figure 14-1. Thus, if

$$v_C = V_m \sin \omega t \tag{14-1}$$

then

$$i_C = I_m \sin(\omega t + 90°) \tag{14-2}$$

where

$$I_m = V_m / X_C \tag{14-3}$$

The quantity X_C is termed *capacitive reactance* and is given by the formula

$$X_C = 1/\omega C \ \Omega \tag{14-4}$$

where $\omega = 2\pi f$ rad/s.

> **Practical Note**
> V_C and I_C are the values that you read on meters, while V_m and I_m are the values that you read on the oscilloscope

Effective (rms) Voltage and Current Relationships: In practice, we usually use effective values rather than peak values. Their ratios are the same however. Thus reactance can also be expressed as

$$X_C = V_C / I_C \ \Omega \tag{14-5}$$

where V_C and I_C are the rms values of v_C and i_C respectively.

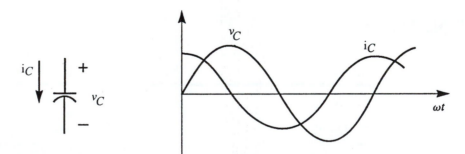

Figure 14-1   Voltage and current for capacitance

MEASUREMENTS

PART A: Phase Relationships for Capacitance

Consider Figure 14-2. Voltage v_S across R_S is in phase with capacitor current i_C. Provided circuit resistance is small compared to X_C, source voltage is approximately equal to the v_C. Thus, the phase dif-

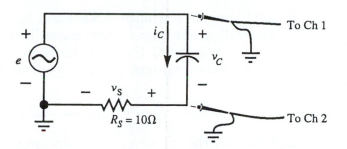

Figure 14-2 Circuit resistance is very small. Therefore, $v_C \approx e$.

ference between Ch1 and Ch2 is approximately equal to the angle between v_C and i_C. It should be close to 90°. Current should lead.

1. a. Assemble the circuit using $R_S = 10\ \Omega$ and $C = 1.0\ \mu F$. Set the oscilloscope for dual channel operation.
   b. Set the signal generator to a 500 Hz sine wave. Sketch the waveforms below, appropriately labeled as v_C and i_C. Determine the phase shift between v_C and i_C and indicate on the diagram. How close is it to the theoretical value of 90 degrees?

   c. Vary the frequency a few hundred Hz up and down and observe the phase shift. Is there any appreciable change? Based on this observation, state in your own words the relationship between voltage across and current through capacitance.

   _____

   _____

PART B: Magnitude Relationships for Capacitance

2. a. Replace R_S of Figure 14-2 with an accurately measured 220-Ω resistor.

   $R_S = $ _____

b. Set the signal generator to 500 Hz at $V_m = 6$ V. With a meter (or an oscilloscope if necessary), carefully measure the voltage V_C and the voltage V_S. Calculate current I_C using Ohm's law and the measured value of V_S. Record as rms values.

$V_C =$ _____ $V_S =$ _____ $I_C =$ _____

c. Calculate the reactance of the capacitor using the measured V_C and I_C.

$X_{C(measured)} = V_C/I_C =$ _____

d. If possible, measure C using a bridge. (Otherwise, use the nominal value.) Compute X_C using Equation 14-4. Compare to the measured value of 2(c).

$C =$ _____ $X_{C(computed)} =$ _____

3. a. Add the second capacitor in series and determine the equivalent reactance of the series combination using the same procedure that you used in Test 2(b) and (c).

$V_C =$ _____ $V_S =$ _____

$I_C =$ _____ $X_{eq(measured)} =$ _____

b. If possible, measure the second capacitor and compute C_{eq} for the series combination. $C_{eq} =$ _____. Use Equation 14-4 to determine $X_{eq(computed)}$. $X_{eq(computed)} =$ _____. Compare to the measured value of 3(a).

4. a. Connect both capacitors in parallel and measure total reactance X_T using the same procedure that you used in Test 2(b) and (c).

$V_C =$ _____ $V_S =$ _____

$I_C =$ _____ $X_{T(measured)} =$ _____

b. Compute C_T for the parallel combination. $C_T =$ _____.

Now use Equation 14-4 to determine $X_{T(computed)}$. $X_{T(computed)} =$

_____. Compare to the measured value of 4(a).

PART C: Variation of Reactance with Frequency

5. a. Replace the parallel combination with a single 1.0-µF capaci-
      tor and measure V_C and V_S and compute reactance (using the
      procedure of Test 2) at each of the frequencies listed in Table
      14-2. Record as $X_{C(measured)}$.
   b. Using Equation 14-4, compute reactance at each of the fre-
      quencies and record as $X_{C(computed)}$.

f (Hz)	V_C	V_S	I_C	$X_{C(measured)}$	$X_{C(computed)}$
100					
200					
300					
400					
500					
600					
700					
800					
900					
1000					

Table 14-2

c. Using the graph sheet of Figure 14-3, plot measured and computed reactances versus frequency. Label each plot.
d. Analyze the results. That is, comment on how well the measured values agree with the computed values.

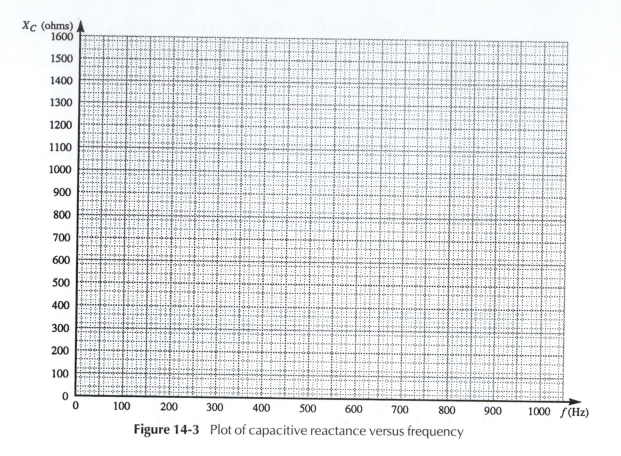

Figure 14-3 Plot of capacitive reactance versus frequency

COMPUTER ANALYSIS

> **Plotting XC with MultiSIM**
> Note that $X_C = V_C/I_C$. If you set $I_C = 1$ A, $X_C = V_C$. Thus, a plot of V_C is also a plot of X_C—i.e., you can read the reactance in ohms directly from the graph.

6. Using either MultiSIM or PSpice, plot the reactance of a 1 μF capacitor from $f = 100$ Hz to 1 kHz. Hint: Plotting reactance using PSpice is straightforward—see Example 16-18 in the text. However, the current version of MultiSIM has no provision for plotting ratios. Thus, another approach is needed. A simple method is to use a 1-amp *ac* current source to drive the capacitor, then plot its resulting voltage—see box. Build the circuit on the screen, then use *Options/Sheet Properties* to determine circuit nodes. Now select *AC Analysis*, type in the desired parameters then Simulate. (Choose linear scales for both x and y to match the diagrams in the book.)

PROBLEMS

7. Consider an ideal capacitor. If you double the capacitance and triple the frequency, what happens to the current if the applied sinusoidal voltage remains the same?

Inductive Reactance

OBJECTIVES

After completing this lab, you will be able to
- measure phase difference between voltage and current in an inductance,
- measure inductive reactance and verify theoretically,
- determine the effect of frequency on inductive reactance.

EQUIPMENT REQUIRED

☐ Dual channel oscilloscope
☐ Signal or function generator
☐ DMM (two)

COMPONENTS

☐ Resistors: 10-Ω, 100-Ω, 1/4-W
☐ Inductors: 2.4-mH, two required. (Hammond Part #1534 or equal. This inductor has a very small resistance which is necessary to approximate an ideal inductor.)

EQUIPMENT USED

Instrument	Manufacturer/Model No.	Serial No.
DMM		
DMM		
Dual channel oscilloscope		
Signal or function generator		

Table 15-1

TEXT REFERENCE

Section 16.5   INDUCTANCE AND SINUSOIDAL AC

DISCUSSION

Inductance and Sinusoidal ac. When an ideal inductance is connected to a sinusoidal voltage source, current lags voltage by 90° as illustrated in Figure 15-1. Thus, if

$$v_L = V_m \sin \omega t \qquad (15\text{-}1)$$

then

$$i_L = I_m \sin(\omega t - 90°) \qquad (15\text{-}2)$$

where

$$I_m = V_m / X_L \qquad (15\text{-}3)$$

The quantity X_L is termed *inductive reactance* and is given by the formula

$$X_L = \omega L \ \Omega \qquad (15\text{-}4)$$

where $\omega = 2\pi f$ rad/s. Inductive reactance represents the opposition that the inductance presents to current and is directly proportional to the product of inductance and frequency. Thus, the higher the frequency, the greater the opposition.

Effective (rms) Voltage and Current Relationships. In practice, we usually use effective values rather than peak values. Their ratios are the same however. Thus reactance can also be expressed as

$$X_L = V_L / I_L \ \Omega \qquad (15\text{-}5)$$

where V_L and I_L are the rms values of v_L and i_L respectively. (Note that V_L and I_L are the values that you read on meters, while V_m and I_m are the values that you read on the oscilloscope.)

Practical Inductors. In reality, inductors have resistance as well as inductance—see Figure 15-2(a). However, if X_L is large compared to

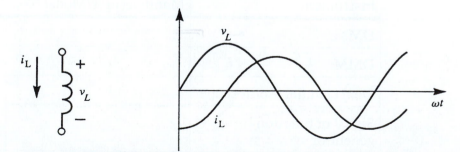

Figure 15-1   Voltage and current for an inductance

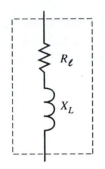

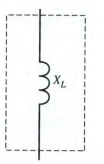

(a) A real inductor

(b) Approximation when R_t has negligible effect.

Figure 15-2   The ideal inductor approximation

R_ℓ , we can neglect resistance and use the ideal model of (b). This means the simple formulas above apply. (In this lab, we will use an inductance where this approximation is very good at higher frequencies. However, at lower frequencies, the error in the approximation starts to show up. You will have a chance to see how good the approximation is at these lower frequencies.)

MEASUREMENTS

PART A: Phase Relationships for Inductance

Consider Figure 15-3. Voltage v_S across R_S is in phase with inductor current i_L. Provided circuit resistance is small compared to X_L, source voltage is approximately equal to the v_L. Thus, the phase difference between Ch1 and Ch2 is approximately equal to the angle between v_L and i_L. It should be close to 90°.

1. a. Measure coil resistance and inductance and the resistance of the sensing resistor then assemble the circuit of Figure 15-3.

$$L = \underline{\hspace{2cm}} \quad R_\ell = \underline{\hspace{2cm}} \quad R_S = \underline{\hspace{2cm}}$$

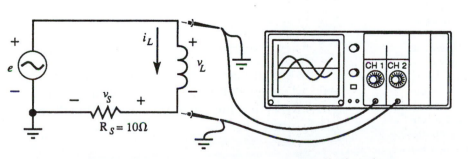

Figure 15-3   Circuit resistance is very small. Therefore, $v_L \approx e$

b. Use a 40 kHz sine wave. Sketch the waveforms below. Label Ch1 as v_L and Ch2 as i_L. Using the procedure of Lab 13, determine the phase shift and indicate it on the diagram. How close is it to the theoretical value of 90°? Vary the frequency a few kHz and observe the shift. Is there any change? Based on this observation, state in your own words the phase relationship between voltage and current for an inductance.

c. The inductor will behave approximately as an ideal inductor provided $X_L \gg R$ where $R = R_\ell + R_s$. Check the approximation at 40 kHz. Use the measured R_ℓ and L determined in Test 1(a).

Practical Notes

1. For this lab, you need to measure voltages with frequencies up to 10 kHz, so ensure that your DMMs can handle this range. If they cannot, use VOMs (as most VOMs can), or if you do not have suitable meters, make measurements with the oscilloscope.

2. For this lab, you need to be able to set frequency quite accurately. However, the frequency dial on many signal generators is not accurately calibrated. If you have such a generator, use your oscilloscope to set frequency.

PART B: Magnitude Relationships for Inductance

2. a. Replace resistor R_S of Figure 15-3 with an accurately measured 100-Ω resistor.

$R_S = $ _____

b. Set the signal generator to a 10-kHz sine wave, 6 V peak to peak. Measure voltage V_L across the inductor and voltage V_S across R_S. Calculate current I_L using Ohm's law and the measured value of V_S.

$V_L = $ _____ $V_S = $ _____ $I_L = $ _____

c. Calculate the reactance of the inductor using the measured V_L and I_L.

$X_{L(measured)} = V_L/I_L = $ _____

d. If possible, measure L using a bridge or RLC meter. (Otherwise, use the nominal value.) Compute X_L using Equation 15-4. Compare to the measured value of Test 2(c).

$L = $ _____ $X_{L(computed)} = $ _____

3. a. Add the second inductor in series and determine total reactance using the same procedure that you used in Test 2(b) and (c).

$V_L = $ _____ $V_S = $ _____

$I_L = $ _____ $X_{T(measured)} = $ _____

b. If possible, measure L for the second inductor and compute L_T for the series combination. $L_T = $ _____. Now use Equation 15-4 to determine $X_{T(computed)}$. $X_{T(computed)} = $ _____. Compare to the measured value of 3(a).

4. a. Connect both inductors in parallel and determine their equivalent reactance using the procedure that you used in Test 2(b) and (c).

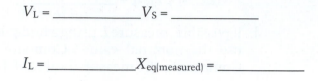

$V_L =$ _____ $V_S =$ _____

$I_L =$ _____ $X_{eq(measured)} =$ _____

   b. Compute L_{eq} for the parallel combination. $L_{eq} =$ _____.

   Now use Equation 15-4 to determine $X_{eq(computed)}$. $X_{eq(computed)} =$

   _____. Compare to the measured value of 4(a).

PART C: Variation of Reactance with Frequency

5. a. Using one of the 2.4-mH inductors, measure V_L and V_S and compute reactance (using the procedure of Test 2) at each of the frequencies listed in Table 15-2. Record as $X_{L(measured)}$.
   b. Using Equation 15-4, compute reactance at each frequency and record as $X_{L(computed)}$.
   c. Using the graph sheet of Figure 15-4, plot measured and computed reactances versus frequency. Label each plot.
   d. The plots should pass through $X_L = 0 \, \Omega$ at $f = 0$ Hz. Do they? (Extrapolate both plots to find out.)
   e. Analyze the results. That is, comment on how well the measured values agree with the computed values. Is the agreement poorer as the frequency gets lower If so, why?

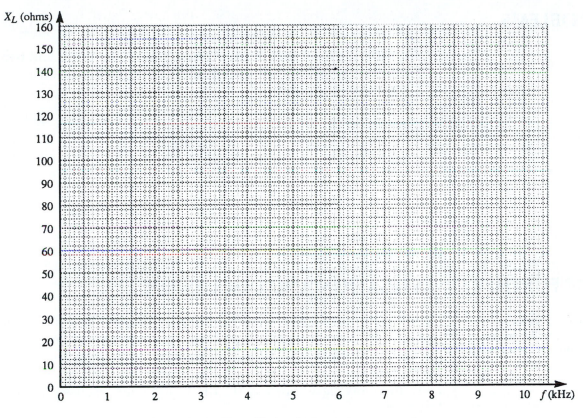

Figure 15-4   Plot of inductive reactance versus frequency

f(kHz)	V_L	V_S	I_L	$X_{L(measured)}$	$X_{L(computed)}$
1					
2					
3					
4					
5					
6					
7					
8					
9					
10					

Table 15-2

PROBLEMS

6. Considering ideal inductors, if you double inductance and triple frequency, what happens to current if the applied sinusoidal voltage remains the same?

FOR FURTHER INVESTIGATION AND DISCUSSION

Analysis Notes
1. Normally frequency plots are made with a log scale for frequency. However, we want to investigate the linearity of impedance versus frequency—thus, use a linear scale for both the horizontal and vertical axes of your plot.
2. MultiSIM users, see note on page 112.

How well do ideal inductors model real inductors? Investigate this question by comparing the impedance characteristic of an ideal inductor to the more realistic model depicted in Section 13.6, Figure 13-21(a) of the text. Assume a 2.4 mH inductor with resistance of 25 Ω and stray capacitance of 0.5 nF. To aid your discussion, do the following.

a. Using MultiSIM or PSpice, compute and plot the magnitude of the impedance (i.e., the reactance) of the ideal inductor from $f = 1$ Hz to 10 kHz.

b. Repeat using the model of Figure 13-21(a). Run 2 sets of plots, one from $f = 1$ Hz to $f = 10$ kHz, and the other from $f = 1$ Hz to $f = 4$ kHz (to better see the low frequency effects).

Write a short report discussing your findings.

NAME _____

DATE _____

CLASS _____

Power in ac Circuits

OBJECTIVES

After completing this lab, you will be able to
- measure power in a single phase circuit,
- verify power relationships,
- verify power factor relationships,
- determine the effect of adding power factor correction.

EQUIPMENT REQUIRED

☐ Single phase wattmeter
☐ ac ammeter
☐ DMM

COMPONENTS

☐ Resistor: 100-Ω, rated 200-W
☐ Capacitor: 30-μF, non-electrolytic, rated for operation at
                120-VAC (1 required)
☐ Inductor: Approximately 0.2-H, rated to handle 2 amps

Safety Note

In this lab, you will be working with 120 VAC. This voltage is dangerous and you must be aware of and observe safety precautions. Familiarize yourself with your laboratory's safety features. **Ensure that power is off when you are assembling, changing or otherwise working on your circuit.**

EQUIPMENT USED

Instrument	Manufacturer/Model No.	Serial No.
Single phase wattmeter		
ac Ammeter		

Table 16-1

TEXT REFERENCE

DISCUSSION

Power in ac Systems. Power to an ac load is given by

$$P = V I \cos \theta \qquad (16\text{-}1)$$

where V and I are the magnitudes of the rms load voltage and current respectively, and θ is the angle between them. (θ is the angle of the load impedance.) The power factor of the load is

$$F_p = \cos \theta \qquad (16\text{-}2)$$

Measuring Power in ac Circuits. Power is measured as in Figure 16-1(a) or (b) with the wattmeter connected so that current passes through its current coil CC and load voltage is applied to its voltage sensing circuit. For this lab, either connection will work.

Power Factor Correction. If a load such as that shown in Figure 16-1 has poor power factor (i.e., θ is large), it will draw excessive current relative to the power transferred to it from the source. One way to improve the power flow in the system is to add power factor correction at the load. Since most power system loads are inductive (because they contain inductive elements such as electric motors, lamp ballasts, and so on), this may be done by adding capacitors in parallel across the load. For loads with a poor power factor, this can dramatically reduce source current. It is important to note however that power factor correction does not change the power requirements of the load—it simply supplies the needed reactive power locally, rather than from the source. However, it greatly reduces source current.

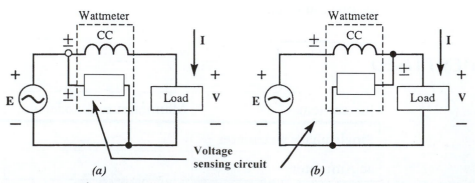

Figure 16-1   Measuring power in an ac circuit. We usually simplify the wattmeter representation as in Figure 16-2.

MEASUREMENTS

If possible, use a variable ac power supply (such as a variable autotransformer) as the source see box.

PART A: Power in a Purely Resistive Circuit

1. Measure the 100-Ω resistor. $R =$ _____. Connect the circuit as in Figure 16-2. (The ammeter may be a standard ac ammeter or the ac current range of a DMM. For the load resistance used here, a 2-A range is adequate. Select a wattmeter to match—e.g., a wattmeter with a 300-W scale.) Carefully measure load voltage, current and power and record in Table 16-2.
2. Using the values of V and R measured in Test 1, compute current, then determine power to the load using each of the formulas $P = V I$, $P = I^2 R$ and $P = V^2/R$. Compare to the value measured with the wattmeter.

> **General Safety Notes**
>
> 1. With power off, assemble your circuit and double check it.
>
> 2. Have your instructor check the circuit before you energize it. If you are using a Powerstat or other variable ac source, gradually increase voltage from zero, watching the meters for signs of trouble.
>
> 3. Turn power off before changing your circuit for the next test.
>
> Check with your instructor for specific safety instructions.

PART B: Power to a Reactive Load

3. a. De-energize the circuit and add 30 µF of capacitance in parallel with the load as in Figure 16-3. Measure load voltage, current, and power and record in Table 16-3.

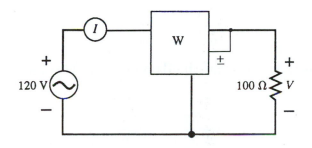

Figure 16-2 Purely resistive load. (The ± voltage connection may be internal to the wattmeter.)

	Measured Values
V	
I	
P	

Table 16-2

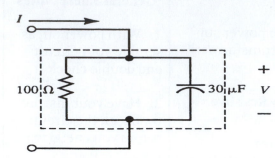

Figure 16-3 Leading power factor load

Measured Values	
V	
I	
P	

Table 16-3

b. Calculate real power using the measured voltage and resistance and compare to the measured value.

c. Calculate reactive power Q

d. Draw the power triangle using the values computed in Tests 3(b) and 3(c). Using the apparent power S from this triangle, compute current. Calculate the percent difference between this and the value measured in Test 3(a). What is the likely source of difference?

e. Determine the power factor from the data of Table 16-3 and Equation 16-1.

f. Determine the power factor from the power triangle of 3(d).

g. Determine the load impedance **Z** from Figure 16-3, then calculate power factor using Equation 16-2.

h. Compare the values of power factor determined in Steps (e), (f), and (g)—i.e., comment on any differences and why they might have occurred.

PART C: Power Factor Correction

Measured Values	
V	
I	
P	

Table 16-4

4. a. Replace the load with the 0.2-H coil, Figure 16-4(a). (Measure its resistance and inductance first.) Measure load voltage, current and power and record in Table 16-4.
    b. Add the 30 µF as in Figure 16-4(b) and observe the decrease in current. Using the techniques of Chapter 17, solve for source current and compare.

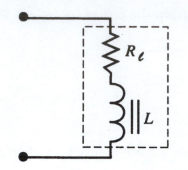
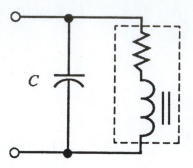

Figure 16-4   (a) Lagging power factor load   (b) Power factor correction

COMPUTER ANALYSIS

Consider Figure 17-22 of the text. Using your calculator, convert the plant load to an equivalent resistance and inductance in series. Omitting the capacitor, set up the resulting circuit using either MultiSIM or PSpice.

a. Insert an ammeter (IPRINT if you are using PSPice) and configure for *ac* operation. Set the source frequency to 60Hz and determine the magnitude of the line current.

b. Add the capacitor across the load and again determine the magnitude of the line current.

c. Change the capacitor to the value that yields unity power factor (*see Practice Problem 4*) and again measure current.

How do the results compare to those in the text?

PROBLEMS

6. A motor delivers 50 hp to a load. The efficiency of the motor is 87% and its power factor is 69%. The source voltage is 277 V, 60 Hz.

a. Determine the power input to the motor.

b. Draw the power triangle for the motor.

c. Determine the source current.

d. Determine how much capacitance is required to correct the power factor to unity.

e. What is the source current when the capacitance of (c) is added?

f. If 50% more capacitance than computed in (c) is added, determine the source current.

NAME _____

DATE _____

CLASS _____

Series ac Circuits

OBJECTIVES

After completing this lab, you will be able to
- calculate current and voltages for a simple series ac circuit,
- measure voltage magnitude and phase angle in a simple series ac circuit,
- verify Kirchhoff's voltage law using measured results,
- measure the internal impedance of a sinusoidal voltage source.

EQUIPMENT REQUIRED

☐ Dual-trace oscilloscope
☐ Signal generator (sinusoidal function generator)
   Note: Record this equipment in Table 17-1.

COMPONENTS

☐ Resistors: 680-Ω (1/4-W carbon, 5% tolerance)
☐ Capacitors: 0.22-μF (10% tolerance)

EQUIPMENT USED

Instrument	Manufacturer/Model No.	Serial No.
Oscilloscope		
Signal Generator		

Table 17-1

TEXT REFERENCE

Section 18.1   OHM'S LAW FOR ac CIRCUITS
Section 18.2   AC SERIES CIRCUITS
Section 18.3   KIRCHHOFF'S VOLTAGE LAW &
                THE VOLTAGE DIVIDER RULE

DISCUSSION

Series ac circuits behave in a manner similar to the operation of dc circuits. Ohm's law, Kirchhoff's current and voltage laws and the various circuit analysis rules apply for ac circuits as well as for dc circuits. The principle differences in these rules and laws apply to ac circuits are outlined as follows:

- All impedances are complex numbers and may be expressed either in rectangular form or polar form (e.g. $\mathbf{Z} = 10\ \Omega + j20\ \Omega = 22.36\ \Omega\angle63.43°$). The real component of the impedance vector represents the resistance while the imaginary component corresponds to the reactance. The imaginary component of the impedance will be positive if the reactance is inductive and negative if the reactance is capacitive.
- *Time-domain values* such as $v_C = 2\sin(\omega t + 30°)$ are converted into *phasor domain* ($\mathbf{V}_C = 1.414\ \text{V}\angle30°$) to permit arithmetic operations using complex numbers.
- Although time domain values are always expressed using peak values, phasor voltages and currents are always expressed with magnitudes in rms (root-mean-square).
- Power calculations are performed using rms values and must consider the *power factor* of the circuit or component.

CALCULATIONS

1. Refer to the circuit of Figure 17-1. Determine the reactance of the capacitor at a frequency of $f = 2$ kHz. Express the circuit impedance $\mathbf{Z}_T$ in both the rectangular form and the polar form.

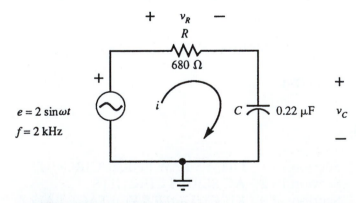

Figure 17-1   Series ac circuit

	X_C	
	$\mathbf{Z}_T$	

2. Convert the time domain form of the voltage source of the circuit of Figure 17-1 into its equivalent phasor domain form. Calculate the phasor current **I** and solve for the phasor voltages $\mathbf{V}_R$ and $\mathbf{V}_C$. Enter your results in Table 17-2.

	E	
	I	
	$\mathbf{V}_R$	
	$\mathbf{V}_C$	

Table 17-2

3. Use complex algebra together with the phasor forms of the voltages (**E**, $\mathbf{V}_R$ and $\mathbf{V}_C$) to verify that Kirchhoff's voltage law applies.

$$\sum \mathbf{V} = 0 = \mathbf{E} - \mathbf{V}_R - \mathbf{V}_C \qquad (17\text{-}1)$$

4. Convert the phasor forms of **I**, $\mathbf{V}_R$, and $\mathbf{V}_C$ into their equivalent time domain values. Enter the results in Table 17-3.

	i	
	V_R	
	V_C	

Table 17-3

MEASUREMENTS

5. Assemble the circuit of Figure 17-1.
6. Refer to the pictorial diagram of Figure 17-2. Connect Ch1 of the oscilloscope to the output of the signal generator. Set the oscilloscope to have an automatic sweep and use Ch1 as the trigger source. Adjust the output of the generator to provide a sinusoidal voltage with an amplitude of 2.0 V (4.0 V_{p-p}) and a frequency of $f = 2$ kHz ($T = 500$ μs).
7. Use Ch2 of the oscilloscope to observe the signal across the capacitor V_C. (You should still have the output of the signal generator displayed on Ch1.) Notice that the voltage across the capacitor is lagging the generator voltage. Adjust the time/division to provide at least one full cycle on the oscilloscope. Accurately sketch and label both the generator voltage e and the capacitor voltage v_C in the space provided on Graph 17-1. In the space below, record the volts/division setting for each channel and the time/division of the oscilloscope.

Ch1 : _____ V/Div.

Ch2 : _____ V/Div.

Time : _____ μs/Div.

8. Use the following expression to calculate the phase angle between capacitor voltage and the generator voltage. (The phase angle is negative since v_C lags e.)

$$\theta = \frac{\Delta t}{T} \times 360° \qquad (17\text{-}2)$$

θ_1	

Figure 17-2 Equipment connection for voltage measurement

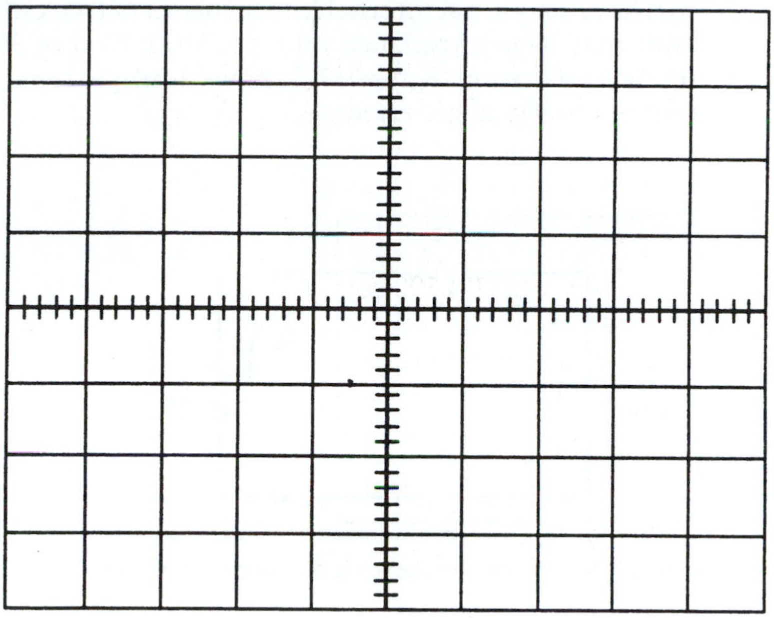

Graph 17-1

9. It is not possible to measure the voltage across the resistor directly since the internal ground connection in the probe of the oscilloscope will result in a short circuit across the capacitor. To overcome this problem, switch the oscilloscope to show the *difference mode* (Ch1 − Ch2), making certain that both Ch1 and Ch2 of the oscilloscope are on the same VOLTS/DIV setting. You should observe a single display, corresponding to the resistor voltage v_R. Since you are using the Ch1 signal as the trigger source, the observed waveform provides the phase shift between the signal generator and the resistor voltage. **Be careful not to adjust the horizontal position. If you accidently move the display, you will need to readjust the display to obtain the waveforms of Step 7**. You should observe that the resistor voltage is leading the generator voltage. Sketch and label the observed resistor voltage v_R as part of Graph 17-1.

10. Calculate the phase shift between the resistor voltage and the generator voltage. Since the resistor voltage is leading the generator voltage, the phase angle is positive.

θ_2	

Internal Impedance of Voltage Sources

All signal generators have some internal impedance which tends to reduce the voltage between the terminals of the signal generator when the generator is under load. Most low-frequency signal generators have a nominal internal impedance of 600 Ω. Other signal generators may have impedance values of 50 Ω, 75 Ω or 300 Ω, depending on the application. Figure 17-3 shows a simple representation of the output of a signal generator.

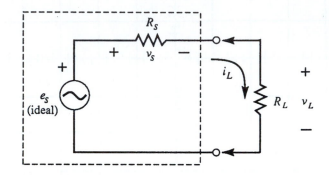

Figure 17-3 Internal impedance of a signal generator

11. Connect the oscilloscope between the output terminals of the signal generator. Adjust the generator to provide a sinusoidal output having an amplitude of 1 V (appearing as 2 $V_{\text{p-p}}$) and a frequency of 1 kHz. Since the signal generator is not loaded, this voltage represents the signal of the ideal voltage source $E_S = 2\ V_{\text{p-p}}$.

12. Connect a 680-Ω load resistor between the output terminals of the signal generator. Reconnect the oscilloscope between the output terminals of the signal generator. You should see that the voltage at the output has now decreased significantly. This represents the loaded output voltage. Measure and record the loaded peak-to-peak output voltage V_L.

V_L	$V_{\text{p-p}}$

CONCLUSIONS

13. Refer to Graph 17-1. Calculate the amplitude V_R of the resistor voltage and the amplitude V_C of the capacitor voltage. From Steps 8 and 10, evaluate the phase angle (with respect to the generator voltage e) for each of these voltages. Express each voltage in its time domain form [e.g. $v_c = V_c \sin(\omega t + \theta_1)$]. Convert the amplitudes into rms quantities and express each voltage in its phasor form. Enter all results in Table 17-4.

Resistor		Capacitor	
V_R	θ	V_C	ρ
$v_R =$		$v_C =$	
$\mathbf{V}_R =$		$\mathbf{V}_C =$	

Table 17-4

14. Compare the measured sinusoidal capacitor voltage v_C of Table 17-4 to the theoretical value of Table 17-2.

_____

_____

15. Compare the measured sinusoidal resistor voltage v_R of Table 17-4 to the theoretical value of Table 17-2.

_____

_____

16. Calculate the actual signal generator resistance using Ohm's law and the measurement of Step 12.

R_S	

FOR FURTHER INVESTIGATION AND DISCUSSION

Use MultiSIM or PSpice to simulate the circuit of Figure 17-1. Measure the amplitude and phase angles of $\mathbf{V}_R$ and $\mathbf{V}_C$. Compare these values to the actual observations. Explain any discrepancies

LAB

18

Parallel ac Circuits

OBJECTIVES

After completing this lab, you will be able to
- measure voltages in a parallel circuit using an oscilloscope,
- use an oscilloscope to indirectly measure current magnitude and phase angles in a simple parallel ac circuit,
- compare measured values to theoretical calculations and verify Kirchhoff's current law,
- determine the power dissipated by a parallel ac circuit.

EQUIPMENT REQUIRED

☐ Dual-trace oscilloscope
☐ Signal generator (sinusoidal function generator)
Note: Record this equipment in Table 18-1.

COMPONENTS

☐ Resistors: 10-Ω (3), 470-Ω (1/4-W carbon, 5% tolerance)
☐ Capacitors: 3300-pF (10% tolerance)
☐ Inductors: 1-mH (iron core, 5% tolerance—Hammond 1534A or equivalent)

EQUIPMENT USED

Instrument	Manufacturer/Model No.	Serial No.
Oscilloscope		
Signal generator		

Table 18-1

TEXT REFERENCE

Section 18.4   AC PARALLEL CIRCUITS
Section 18-5   KIRCHHOFF'S CURRENT LAW & THE CURRENT DIVIDER RULE

DISCUSSION

The equivalent impedance of a parallel circuit is determined by finding the sum of the admittances of all branches. The important point to remember when calculating total admittance is that all admittances are expressed as complex values. This means that we must use complex algebra to find the solution, which is also a complex number.

Refer to the circuit of Figure 18-1. The equivalent admittance of the circuit is determined as

$$\mathbf{Y}_T = \frac{1}{R} + j\frac{1}{X_C} - j\frac{1}{X_L} \tag{18-1}$$

This gives an equivalent circuit impedance of

$$\mathbf{Z}_T = \frac{1}{\mathbf{Y}_T} = \frac{1}{\dfrac{1}{R} + j\dfrac{1}{X_C} - j\dfrac{1}{X_L}} \tag{18-2}$$

In order to further analyze the circuit, it is necessary to convert the ac voltage source into its equivalent phasor form. The circuit current **I** is then easily determined by applying Ohm's law. The current through each component in the circuit is similarly found by applying Ohm's law to each branch or by applying the current divider rule as follows:

$$\mathbf{I}_x = \frac{\mathbf{E}}{\mathbf{Z}_x} = \frac{\mathbf{Z}_T}{\mathbf{Z}_x}\mathbf{I} \tag{18-3}$$

Regardless of the method used to determine currents, Kirchhoff's current law must apply at any node in the circuit. Therefore,

$$\sum \mathbf{I} = 0 \tag{18-4}$$

where each current is in its phasor form.

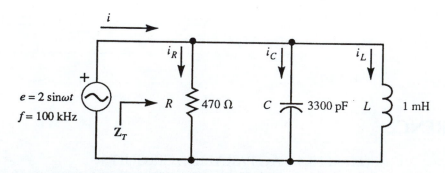

Figure 18-1   Parallel ac circuit

CALCULATIONS

1. Refer to the circuit of Figure 18-1. Determine the reactances of the capacitor and the inductor at a frequency of $f = 100$ kHz. Calculate the circuit impedance $\mathbf{Z}_T$ and express the result in both rectangular and polar form. Enter the data in Table 18-2.

X_C	
X_L	
$\mathbf{Z}_T$	

Table 18-2

2. Convert the time domain form of the voltage source of the circuit of Figure 18-1 into its equivalent phasor domain form. Calculate the phasor current $\mathbf{I}$ and solve for the phasor currents $\mathbf{I}_R$, $\mathbf{I}_C$, and $\mathbf{I}_L$. Enter your results in Table 18-3.

E	
I	
$\mathbf{I}_R$	
$\mathbf{I}_C$	
$\mathbf{I}_L$	

Table 18-3

3. Use complex algebra together with the phasor currents $\mathbf{I}_R$, $\mathbf{I}_C$, and $\mathbf{I}_L$ to verify that Kirchhoff's current law applies at node a.

$$\mathbf{I} = \mathbf{I}_R + \mathbf{I}_C + \mathbf{I}_L \qquad (18\text{-}5)$$

4. Convert the phasor currents $\mathbf{I}$, $\mathbf{I}_R$, $\mathbf{I}_C$, and $\mathbf{I}_L$ into their equivalent time domain forms. Enter the results in Table 18-4.

i	
i_R	
i_C	
i_L	

Table 18-4

Current cannot be measured directly with an oscilloscope. However, by strategically placing small series *sensing resistors* into a circuit and then measuring voltage across these resistors, current through each branch is found by applying Ohm's law.

MEASUREMENTS

5. Assemble the circuit of Figure 18-2. Notice that three 10-Ω *sensing resistors* have been added into the circuit to help in determining branch currents. Since these resistors are small in comparison to the impedance in the branch, they will not significantly load the circuit and their effects may be ignored.
6. Connect Ch1 of the oscilloscope to the output of the signal generator at point *a*. Set the oscilloscope to have an automatic sweep and use Ch1 as the *trigger source*. Adjust the output of the generator to provide a sinusoidal voltage with an amplitude of 2.0 V (4.0 V$_{p-p}$) and a frequency of $f = 100$ kHz ($T = 10$ μs).
7. Use Ch2 of the oscilloscope to observe the voltage at point *b*. This is the voltage across sensing resistor R_1. Measure the observed peak-to-peak voltage V_1. Measure the phase angle $_1$ of the voltage V_1 with respect to the generator voltage. Record your results in Table 18-5.

   Since the observed waveform is across the 10-Ω resistor, the amplitude of the current *i* is now easily calculated from the peak-to-peak voltage as

$$I = \frac{V_1/2}{10\ \Omega}$$

   (18-6)

   Write the time-domain expression for *i* using your measurements and calculations. Enter your results in Table 18-5.

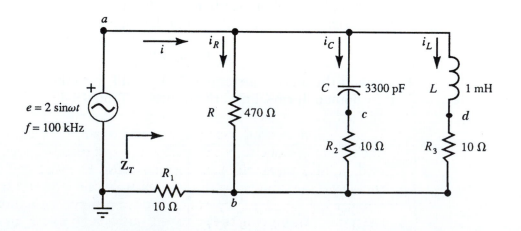

Figure 18-2   Using sensing resistors for current measurement

V_1	$V_{p\text{-}p}$
θ_1	
i	

Table 18-5

8. Since we no longer need R_1 in the circuit, it may be removed and replaced with a short circuit. Measure the peak-to-peak voltage V_R and calculate current i_R using Ohm's law. Use your measurements and calculations to record the time-domain expression for i_R in Table 18-6.

V_R	$V_{p\text{-}p}$
θ	
i_R	

Table 18-6

9. Ensure that Ch1 of the oscilloscope is connected to point a (generator voltage) and that the sensing resistor, R_1 is removed from the circuit.
    Connect Ch2 of the oscilloscope to point c to measure the voltage across the sensing resistor R_2. Record the peak-to-peak voltage V_2. Measure and record the phase angle between the generator voltage and the sensing voltage V_2. You should observe that the voltage V_2 (and hence, the current) is leading the generator voltage. Use your measurements and Ohm's law to determine the sinusoidal expression for i_c. Record the result in Table 18-7.

V_1	$V_{p\text{-}p}$
θ_2	
i_C	

Table 18-7

10. With Ch1 of the oscilloscope connected to point a, connect Ch2 of the oscilloscope to point c. Measure and record the peak-to-peak voltage across the sensing resistor R_3. Measure and record the phase angle between the generator voltage and the sensing voltage V_2. You should observe that the voltage V_3 (and hence, the current) is lagging the generator voltage. Use your measurements and Ohm's law to determine the sinusoidal expression for i_L. Record the result in Table 18-8.

V_3	$V_{p\text{-}p}$
θ_3	
i_L	

Table 18-8

CONCLUSIONS

11. Compare the measured currents i, i_R, i_C, and i_L of Table 18-5 through Table 18-8 to the theoretical values of Table 18-4.

_____

_____

_____

12. Convert each of the measured currents i, i_R, i_C, and i_L into its phasor form. Enter the results in Table 18-9.

I	
$\mathbf{I}_R$	
$\mathbf{I}_C$	
$\mathbf{I}_L$	

Table 18-9

13. Use complex algebra to show that the data of Table 18-9 verifies Kirchhoff's current law.

14. Use the measured currents $\mathbf{I}$, $\mathbf{I}_R$, $\mathbf{I}_C$, and $\mathbf{I}_L$ to calcultate the total power dissipated by the circuit in Figure 18-2. (You should observe that the sensing resistors dissipate very little power.)

P_T	

Series-Parallel ac Circuits

OBJECTIVES

After completing this lab, you will be able to

- analyze a series-parallel circuit to determine the current through and voltage across each element in a series-parallel circuit,
- measure voltage across each element in a series-parallel circuit using an oscilloscope and use the measurements to determine the current through each element of a series-parallel circuit,
- calculate the power dissipated by each element in a circuit,
- use measurements to verify that the actual powers dissipated correspond to theory.

EQUIPMENT REQUIRED

☐ Dual trace oscilloscope
☐ Signal generator (sinusoidal function generator)
   Note: Record this equipment in Table 19-1.

COMPONENTS

☐ Resistors: 10-Ω (2), 470-Ω (1/4-W carbon, 5% tolerance)
☐ Capacitors: 3300-pF (10% tolerance)
☐ Inductors: 1-mH (iron core, 5% tolerance)

EQUIPMENT USED

Instrument	Manufacturer/Model No.	Serial No.
Oscilloscope		
Signal generator		

Table 19-1

TEXT REFERENCE

Section 18-6   SERIES-PARALLEL CIRCUITS

DISCUSSION

The equivalent impedance of a series-parallel ac circuit is determined in a manner which is similar to that used in finding the equivalent resistance of a series-parallel circuit, with the exception that vector algebra is used in determining the total impedance at a given frequency. It is necessary to decide which elements or branches are in series and which are in parallel. The resultant impedance is the combination of the various connections. Once we have the total impedance, it is a simple matter to calculate the total current. By applying appropriate circuit theory, the current, voltage, and power of the various components of the circuit may then be found.

Refer to the circuit of Figure 19-1. Notice that resistor R and inductor L are in parallel. This parallel connection is then seen to be in series with capacitor C. The total impedance of the circuit is therefore determined as

$$\mathbf{Z}_T = -jX_C + R \parallel jX_L \tag{19-1}$$

The power provided to the circuit by the voltage source is calculated as

$$P_T = EI\cos\theta = \frac{E^2}{Z_T}\cos\theta \tag{19-2}$$

In the above expression, E and I are the rms values of the sinusoidal voltage e and current i. Z_T is the magnitude of the circuit impedance and   is the angle between the current phasor $\mathbf{I} = \mathbf{I}_C$ and the

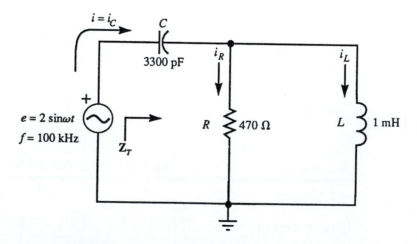

Figure 19-1   Series-parallel circuit

voltage phasor **E**. (This is the same angle as the angle in the impedance vector $\mathbf{Z}_T$.)

On further examination of Figure 19-1, we see that only the resistor can dissipate power. (Inductors and capacitors do not dissipate
power.) This means that the total power in the circuit must also be
the same as the power dissipated by the resistor, namely

$$P_R = \frac{V_R^2}{R} = I_R^2 R \qquad (19\text{-}3)$$

where V_R and I_R are rms quantities.

CALCULATIONS

1. Refer to the circuit of Figure 19-1. Determine the reactances of
   the capacitor and the inductor at a frequency of $f = 100$ kHz. Calculate the circuit impedance $\mathbf{Z}_T$. Enter all values here.

X_C	
X_L	
$\mathbf{Z}_T$	

2. Convert the time-domain form of the voltage source into its
   equivalent phasor-domain form. Calculate the phasor current
   through each element of the circuit. Enter your results in Table
   19-2.

E	
$\mathbf{I}_R$	
$\mathbf{I}_C$	
$\mathbf{I}_L$	

Table 19-2

3. Calculate and record the total power provided to the circuit by the voltage source.

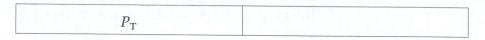

P_T	

4. Use complex algebra to show that currents $\mathbf{I}_R$, $\mathbf{I}_C$, and $\mathbf{I}_L$ satisfy Kirchhoff's current law

$$\sum \mathbf{I} = 0$$

(19-4)

MEASUREMENTS

5. Assemble the circuit of Figure 19-2. Notice that two 10-Ω *sensing resistors* have been added to the circuit to help in determining branch currents. Since these resistors are small in comparison to the impedance in the branch, they will not significantly load the circuit.

6. Connect Ch1 of the oscilloscope to the output of the signal generator at point a. Set the oscilloscope to an automatic sweep and use Ch1 as the *trigger source*. Adjust the output of the generator to provide a sinusoidal voltage with an amplitude of 2.0 V (4.0 V_{p-p}) at a frequency of $f = 100$ kHz ($T = 10$ μs).

7. Use Ch2 of the oscilloscope to observe the voltage at point c. This is the voltage across sensing resistor R_1. Measure the peak-to-peak voltage V_1. Determine the phase angle θ_1 of voltage v_1 with respect to the generator voltage e. Write the time-domain expression for i using your measurements and calculations. Record your results in Table 19-3.

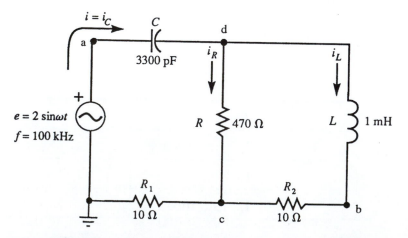

Figure 19-2 Using sensing resistors for current measurement

V_1	$V_{\text{p-p}}$
θ_1	
$i = i_c$	

Table 19-3

8. Remove resistor R_1 from the circuit and replace it with a short circuit. Move Ch2 of the oscilloscope to observe the voltage at point d (the voltage across resistor R). Measure the peak-to-peak voltage V_R and determine phase angle θ_R of voltage v_R with respect to the generator voltage e. Record your results in Table 19-4. Determine the amplitude i_R and enter the time-domain expression for i_R in Table 19-4.

V_R	$V_{\text{p-p}}$
θ_R	
i_R	

Table 19-4

9. With Ch1 of the oscilloscope connected to point a, connect Ch2 of the oscilloscope to point b. Measure and record the peak-to-peak voltage across the sensing resistor R_2. Measure and record the phase angle between the generator voltage and the sensing voltage V_2. Use your measurements and Ohm's law to determine the sinusoidal expression for i_L. Record the result in Table 19-5.

V_2	$V_{\text{p-p}}$
θ_2	
i_L	

Table 19-5

CONCLUSIONS

10. Convert the measured currents i_R, i_C, and i_L of Table 19-3 through Table 19-5 into their phasor forms. Enter your results in Table 19-6 and compare them with the theoretical values of Table 19-2.

$I = I_C$	
I_R	
I_L	

Table 19-6

_____

_____

11. Use complex algebra to verify that the data of Table 19-6 satisfy Kirchhoff's current law.

12. Use the phasors **E** and **I** to calculate the total power delivered to the circuit by the voltage source. Enter your calculation in the space provided.

P_T	

13. Now use the rms values of V_R and I_R to determine the power dissipated by the resistor. Enter the result in the space below. How does this value compare with the total power delivered to the circuit by the voltage source? Offer an explanation for any variation.

P_R	

_____

_____

PROBLEMS

14. Refer to the circuit of Figure 19-1. If the frequency of the signal generator is increased to 200 kHz, determine the following:
   a. Total impedance Z_T,
   b. Currents **I**, I_R, and I_L,
   c. Power P_R dissipated by the resistor,
   d. Power P_T deliverd by the voltage source.
15. Repeat Problem 14 if the frequency of the generator is decreased to 50 kHz.

Thévenin's and Norton's Theorems (ac)

OBJECTIVES

After completing this lab, you will be able to

- calculate the Thévenin and Norton equivalents of an ac circuit,
- measure the Thévenin (open circuit) voltage and the Norton (short circuit) current of an ac circuit,
- calculate the Thévenin impedance of a circuit using the measured values of Thévenin voltage and Norton current,
- measure the load impedance which results in a maximum transfer of power to the load.

EQUIPMENT REQUIRED

☐ Dual trace oscilloscope
☐ Signal generator (sinusoidal function generator)
☐ DMM
Note: Record this equipment in Table 20-1.

COMPONENTS

☐ Resistors: 10-Ω, 1.5-kΩ (1/4-W carbon, 5% tolerance)
              5-kΩ variable resistor
☐ Capacitors: 2200-pF (10% tolerance)
☐ Inductors: 2.4-mH (iron core, 5% tolerance)

EQUIPMENT USED

Instrument	Manufacturer/Model No.	Serial No.
Oscilloscope		
Signal generator		
DMM		

Table 20-1

TEXT REFERENCE

Section 20-3   THÉVENIN'S THEOREM—INDEPENDENT
SOURCES
Section 20-4   NORTON'S THEOREM—INDEPENDENT SOURCES
Section 20-6   MAXIMUM POWER TRANSFER THEOREM

DISCUSSION

Thévenin's theorem allows us to convert any two-terminal linear bilateral circuit into an equivalent circuit consisting of a voltage source $\mathbf{E}_{Th}$ in series with an impedance $\mathbf{Z}_{Th}$ as illustrated in Figure 24-1(a). When a load is connected across the two terminals of the circuit, the current through (or voltage across) the load is easily calculated by analyzing the equivalent circuit. The Thévenin voltage of an equivalent circuit is determined by removing the load from the circuit and measuring the open circuit voltage.

Norton's theorem is the duality of Thévenin's theorem in that it converts any two-terminal linear bilateral circuit into an equivalent circuit consisting of a current source $\mathbf{I}_N$ in parallel with an impedance $\mathbf{Z}_N$ as shown in Figure 20-1(b). The Norton current of an equivalent circuit is determined by replacing the load with a short circuit and measuring the current through the load.

Unlike dc circuits, the Thévenin (and Norton) impedance of an ac circuit cannot be measured directly. Rather, the impedance is determined indirectly by using the equalence between the Thévenin and Norton circuits. Since the circuits are equivalent, the following relationship must apply:

$$\mathbf{Z}_{Th} = \mathbf{Z}_N = \frac{\mathbf{E}_{Th}}{\mathbf{I}_N}$$

(20-1)

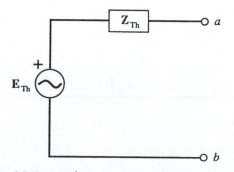

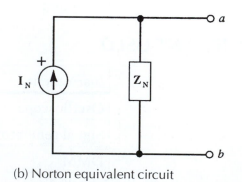

Figure 20-1   (a) Thévenin equivalent circuit   (b) Norton equivalent circuit

CALCULATIONS

1. Refer to the circuit of Figure 20-2. Determine the reactances of the capacitor and the inductor at a frequency of $f = 100$ kHz. Enter the results here.

X_C	
X_L	

2. Convert the time-domain form of the voltage source in Figure 20-2 into its equivalent phasor-domain form and enter the result here.

E	

3. Determine the Thévenin equivalent circuit to the left of terminals a and b in the circuit of Figure 20-2. Sketch the equivalent circuit in this space.

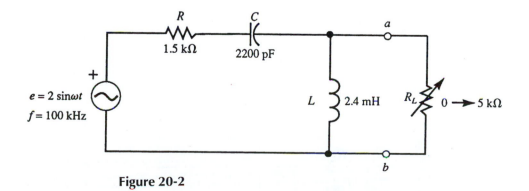

Figure 20-2

4. Determine the Norton equivalent circuit to the left of terminals *a* and *b* in the circuit of Figure 20-2. Sketch the equivalent circuit in this space.

5. *Absolute maximum power* will be delivered to the load impedance when the load is the complex conjugate of the Thévenin (or Norton) impedance. For what value of load impedance Z_L will the circuit of Figure 20-2 transfer maximum power to the load? Solve for the absolute maximum load power.

Z_L	
P_{max}	

6. The load in the circuit of Figure 20-2 does not have any reactance. In this case, absolute maximum power cannot be transferred to the load. *Relative maximum power* will be transferred to the load impedance when the value of load resistance is equal to the following:

$$R = \sqrt{R_{Th}^2 + X_{Th}^2}$$

(20-2)

For what value of load resistance R_L will the circuit of Figure 20-2 transfer maximum power to the load? Solve for the relative maximum load power and enter the results in the space provided.

R_L	
P_{max}	

MEASUREMENTS

7. Assemble the circuit of Figure 20-2, temporarily omitting the load resistor R_L. Connect Ch1 of the oscilloscope to the output of the signal generator and adjust the generator to provide an output of 2.0 V_p (4.0 $V_{p\text{-}p}$) at f = 100 kHz.

8. While using Ch1 as the trigger source of the oscilloscope, connect Ch2 between terminals a and b of the circuit. Measure and record the peak-to-peak value of the open circuit voltage amplitude E_{Th} and the phase shift. (The phase shift is measured with respect to the generator voltage observed on Ch1.)

E_{Th}	$V_{p\text{-}p}$
P_{max}	

9. Place a 10-Ω *sensing resistor* between terminals a and b. The sensing resistor has a very low impedance with respect to the other components in the circuit and so has a minimal loading effect. Readjust the generator voltage to ensure that the output is 2.0 V_p. Measure the peak-to-peak voltage across the sensing resistor and determine the peak-to-peak value of the "short circuit" current I_N and the phase shift θ.

V_{ab}	$V_{p\text{-}p}$
I_N	$mA_{p\text{-}p}$
θ	

10. Use the DMM to adjust the 5-kΩ variable resistor for a resistance of 500 Ω. Remove the sensing resistor and insert the variable resistor between terminals a and b of the circuit. Adjust the supply voltage for 2.0 $V_{p\text{-}p}$ and measure the amplitude of the output voltage, V_L. Enter your measurement in Table 20-2.

11. Remove R_L from the circuit and incrementally increase the resistance by 500 Ω. Repeat Step 10. Keep increasing the load resistance until R_L = 5 kΩ. Enter all data in Table 20-2.

R_L	V_L
500 Ω	
1000 Ω	
1500 Ω	
2000 Ω	
2500 Ω	
3000 Ω	
3500 Ω	
4000 Ω	
4500 Ω	
5000 Ω	

Table 20-2

CONCLUSIONS

12. Convert the measured voltage and phase angle for the Thévenin voltage of Step 8 into its correct phasor form. Record your result in the space provided below. How does this value compare to the theoretical value determined in Step 3?

$\mathbf{E}_{Th}$	

13. Convert the measured current and phase angle for the Norton current of Step 9 into its correct phasor form. Enter the result below. How does this value compare to the theoretical value determined in Step 4?

$\mathbf{I}_N$	

14. Use the phasors of Steps 12 and 13 to calculate the Thévenin

(and Norton) impedance. How does this value compare to the theoretical value determined in Steps 3 and 4?

$Z_{Th} = Z_N$	

15. Use the data of Table 20-2 to calculate the power delivered to the load for each of the resistor values. (Remember that you will need to convert each voltage measurement into its equivalent rms value in order to calculate the power.) Enter the results in Table 20-3.

R_L	P_L
500 Ω	
1000 Ω	
1500 Ω	
2000 Ω	
2500 Ω	
3000 Ω	
3500 Ω	
4000 Ω	
4500 Ω	
5000 Ω	

Table 20-3

16. Use the data of Table 20-2 to sketch a graph of power (in microwatts) versus load resistance (in ohms). Connect the points with the best smooth continuous curve. (A correctly drawn curve will not be drawn from point-to-point.)
17. Use the curve of Graph 20-1 to determine the approximate value of load resistance R_L for which the load receives maximum power from the circuit. Enter the result here. How does this value compare to the value determined in Step 6?

R_L	

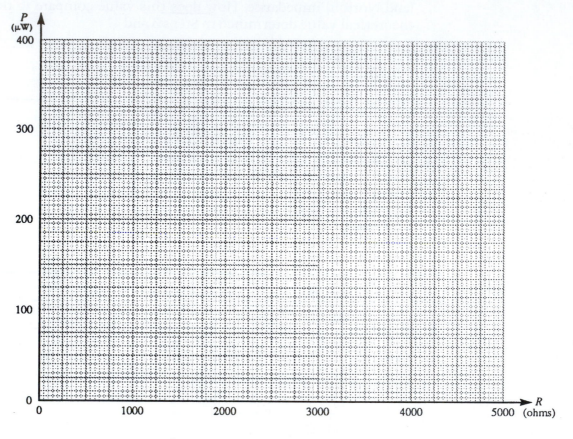

Graph 20-1

PROBLEMS

18. If the circuit of Figure 20-2 (f = 100 kHz) is to provide absolute maximum power to the load, what value of capacitance (in μ or inductance (in mH) must be added in series with the load resistance?

19. Determine the Norton equivalent of the circuit to the left of points a and b in the circuit of Figure 20-2. Assume that the circuit operates at f = 200 kHz.

20. Refer to the circuit of Figure 20-2.
    a. Determine the Thévenin equivalent of the circuit to the left of points a and b, assuming that the circuit operates at a frequency of f = 50 kHz.
    b. Solve for the load impedance which will result in a maximum transfer of power to the load.
    c. Calculate the maximum power which can be transferred to the load.

Series Resonance

OBJECTIVES

After completing this lab, you will be able to
- calculate the resonant frequency of a series resonant circuit,
- solve for the maximum output voltage of a resonant circuit using the *quality factor Q* of the circuit,
- measure the bandwidth of a series resonant circuit,
- measure the impedance at frequencies above and below the resonant frequency and observe that it is purely resistive only at resonance,
- sketch the circuit current as a function of frequency and explain why the response has a bell-shaped curve when plotted on a semi-logarithmic graph.

EQUIPMENT REQUIRED

☐ Dual trace oscilloscope
☐ Signal generator (sinusoidal function generator)
☐ DMM
Note: Record this equipment in Table 21-1.

COMPONENTS

☐ Resistors: 15-Ω (1/4-W carbon, 5% tolerance)
☐ Capacitors: 0.33-μF (10% tolerance)
☐ Inductors: 2.4-mH (iron core, 5% tolerance)

EQUIPMENT USED

Instrument	Manufacturer/Model No.	Serial No.
Oscilloscope		
Signal generator		
DMM		

Table 21-1

TEXT REFERENCE

Section 21-1   SERIES RESONANCE
Section 21-2   QUALITY FACTOR, Q
Section 21-3   IMPEDANCE OF A SERIES RESONANT CIRCUIT
Section 21-4   POWER, BANDWIDTH AND SELECTIVITY
OF A SERIES RESONANT CIRCUIT

DISCUSSION

Resonant circuits are used throughout electronics as a means of passing a range of frequencies, while rejecting all other frequencies. These circuits have important applications in communications where they are used in circuits such as receivers to tune into a particular station or channel. Figure 21-1 represents a typical series resonant circuit.

At the resonant frequency, the reactance of the inductor is exactly equal to the reactance of the capacitor. Since they are equal with opposite phase, the reactances cancel and the total impedance of the circuit is purely resistive at resonance. The resonant frequency (in hertz) of a series circuit is given as

$$f_S = \frac{1}{2\pi\sqrt{LC}} \tag{21-1}$$

At the resonant frequency, the current (and power) in the circuit is maximum, resulting in a maximum output voltage appearing across the inductor. Since the reactance of the inductor can be many times greater than the resistance of the circuit, the output voltage may be many times greater than the applied signal. This characteristic is one of the advantages of using a resonant circuit since the out-

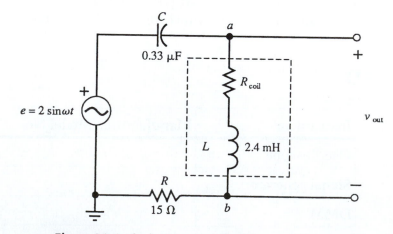

Figure 21-1   Series resonant circuit

put voltage is amplified without the need for *active components* such as transistors.

The *quality factor Q* of a resonant circuit is defined as the ratio of reactive power to the real power at the resonant frequency. It can be shown that the quality factor for the circuit of Figure 21-1 is determined as

$$Q = \frac{X_L}{R} = \frac{\omega L}{R_S + R_{\text{coil}}} \qquad (21\text{-}2)$$

The Q of a circuit is used to determine the range of frequencies which will be passed by a given resonant circuit. If the circuit has a high Q (greater than 10), it will pass a narrow range of frequencies and the circuit is said to have a *high selectivity*. Conversely, if the Q of the circuit is small, the circuit will pass a broader range of frequencies and the circuit is said to have a *low selectivity*. The *bandwidth (BW)* of a resonant circuit is defined as the difference between the the half-power frequencies, namely the frequencies at which the circuit dissipates half the power that would be dissipated at resonance. The bandwidth (in hertz) of a resonant circuit is determined as

$$BW = \frac{f_S}{Q} \qquad (21\text{-}3)$$

The half-power frequencies occur on either side of the resonant frequency. If the quality factor of the circuit is large (Q 10) then the half-power frequencies are given approximately as

$$f_1 \cong f_S - \frac{BW}{2} \qquad (21\text{-}4)$$

and

$$f_2 \cong f_S + \frac{BW}{2} \qquad (21\text{-}5)$$

CALCULATIONS

1. Determine the resonant frequency for the circuit of Figure 21-1.

f_S	

2. Prior to starting the lab, obtain an inductor from your lab instructor. Use the DMM ohmmeter to measure the dc resistance of the inductor. Enter the result below.

R_{coil}	

3. Using the measured resistance of the inductor, calculate the phasor form of current **I** at resonance. Enter the result in the space provided below.

I	

4. Determine the phasor form of output voltage **V**$_{out}$ appearing across the inductor. Enter the result below. You will need to consider the effect of R_{coil}.

V$_{out}$	

5. Calculate the quality factor Q, bandwidth BW, and approximate half-power frequenies f_1 and f_2 for the circuit. Record the results in the space provided.

Q	
BW	
f_1	
f_2	

MEASUREMENTS

6. Assemble the circuit shown in Figure 21-1. Connect Ch1 of the oscilloscope to the output of the signal generator and adjust the generator to provide an amplitude of 2.0 V$_p$ (4.0 V$_{p-p}$) at a frequency of $f = 1$ kHz.

7. Connect Ch2 of the oscilloscope across the resistor R and measure the amplitude of V_R. (In order to simplify this measurement, it is generally easier to measure the peak-to-peak voltage and then divide by two.) Use the measured voltage to calculate the amplitude of the current I. Enter the results in Table 21-2.

8. Increase the frequency of the signal generator to the frequencies indicated in Table 21-2. Due to loading effects, the amplitude of the generator will tend to drift. Ensure that the output of the signal generator is kept constant at 2.0 V$_p$. Measure the amplitude of V_R for each frequency and calculate the corresponding amplitude of current I. Enter the values in Table 21-2.

f	V_R	I
1 kHz		
2 kHz		
3 kHz		
4 kHz		
4.5 kHz		
5 kHz		
5.5 kHz		
6 kHz		
6.5 kHz		
7 kHz		
8 kHz		
9 kHz		
10 kHz		

Table 21-2

9. For which frequency f in Table 21-2 is the circuit current a maximum? Adjust the generator to provide an output of 2.0 V_p at this frequency. While observing the oscilloscope, adjust the the generator frequency until the resistor voltage V_R is at the maximum value. Record the resonant frequency f_s and the corresponding resistor voltage V_R in Table 21-3. Calculate the current at resonance.

10. Decrease the frequency until the output voltage V_R is reduced to 0.707 of the maximum value found in Step 9. Record the lower half-power frequency f_1 and the corresponding resistor voltage V_R in Table 21-3. (Ensure that the output of the signal generator is at 2.0 V_p.)

    Increase the frequency above the resonant frequency until the output voltage V_R is again reduced to 0.707 of the maximum value found in Step 9. Record the upper half-power frequency f_2 and the corresponding resistor voltage V_R in Table 21-3. (Ensure that the output of the signal generator is at 2.0 V_p.)

    Calculate the current for each frequency.

f	V_R	I
$f_1 =$		
$f_S =$		
$f_2 =$		

Table 21-3

11. Set the signal generator to the resonant frequency determined in Step 9 and adjust the amplitude for 2.0 V_p. With Ch1 of the oscilloscope at the output of the signal generator, use Ch2 to measure the voltage across resistor R. Calculate the phasor form of current **I** at resonance. (You should observe that v_R and e are in phase.) Record your results here.

V_R	
θ	
I	

12. Adjust the signal generator for a frequency $f = f_1$ and a voltage of 2.0 V_p. Measure the magnitude and phase angle of the voltage V_R. Calculate the phasor form of current **I** at this frequency.

V_R	
θ_1	
$\mathbf{I}_1$	

13. Adjust the signal generator for a frequency $f = f_2$ and a voltage of 2.0 V_p. Measure the magnitude and phase angle of the voltage V_R. Calculate the phasor form of the current **I** at this frequency.

V_R	
θ_2	
$\mathbf{I}_2$	

14. Set the signal generator to the resonant frequency determined in Step 9 and adjust the amplitude for 2.0 V_p. Place Ch1 of the oscilloscope at terminal a of the circuit and Ch2 at terminal b. Use the difference mode of the oscilloscope to measure the amplitude of the output voltage. You should observe that the amplitude of this voltage is larger than the amplitude at the output of the signal generator. Record the amplitude of output voltage V_{out} in the space provided.

V_{out}	

CONCLUSIONS

15. Plot the data of Tables 21-2 and 21-3 on the semi-logaritmic scale of Graph 21-1. Connect the points with the best smooth continuous curve.

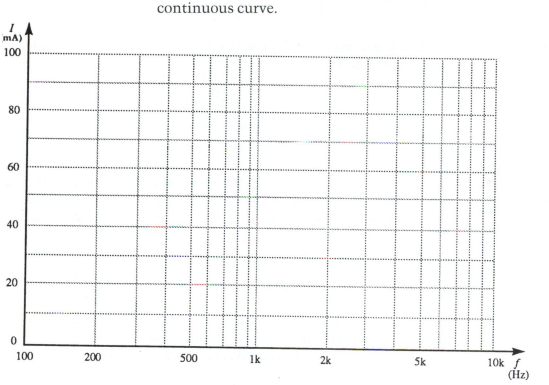

Graph 21-1

16. Compare the measured resonant frequency f_s found in Step 9 to the theoretical frequency determined in Step 1.

_____

_____

17. Compare the measured half-power frequencies from Step 10 to the theoretical values determined in Step 5.

_____

_____

_____

_____

18. Use the measured half-power frequencies to calculate the bandwidth of the circuit. Enter the results below.

$$BW = f_2 - f_1 \qquad\qquad (21\text{-}6)$$

BW	

19. Calculate the Q of the circuit and enter the result in the space provided.

$$Q = \frac{f_S}{BW} \qquad\qquad (21\text{-}7)$$

Q	

20. Use the data of Step 11 to compare the phase angle of the current with respect to the signal generator at resonance. Based on this result, is the circuit resistive, inductive, or capacitive when $f = f_S$?

_____

_____

21. Use the data of Step 12 to compare the phase angle of the current with respect to the signal generator when $f = f_1$. Based on this result, is the circuit resistive, inductive, or capacitive when $f < f_S$?

_____

_____

22. Use the data of Step 13 to compare the phase angle of the current with respect to the signal generator when $f = f_2$. Based on this result, is the circuit resistive, inductive, or capacitive when $f > f_S$?

_____

_____

23. In Step 14 , you should have observed that the output voltage of the circuit at resonance is larger than the applied signal generator voltage. Calculate and record the ratio of the amplitudes V_{out}/E. Since the output voltage is taken across the inductor, you should observe that this ratio is very close to the expected Q of the circuit. Compare this result with that obtained in Step 19.

$Q = V_{out}/E$	

_____

_____

FOR FURTHER INVESTIGATION AND DISCUSSION

24. If the resistance of the inductor R_{coil} was higher than the measured value, what would happen to f_S, Q, BW, and the output voltage v_{out} at resonance?

    f_S:

_____

_____

    Q:

_____

_____

    BW:

_____

_____

    v_{out} at resonance:

_____

_____

25. Use MultiSIM or PSpice to simulate the circuit of Figure 21-1. Use a coil resistance of $R_{coil} = 5\ \Omega$. Find the resonant frequency, bandwidth, and the quality factor of the circuit.
26. Repeat Step 25 by letting $R_{coil} = 10\ \Omega$.

Parallel Resonance

OBJECTIVES

After completing this lab, you will be able to
- calculate the resonant frequency of a parallel resonant circuit,
- solve for the maximum output voltage of a parallel resonant circuit,
- measure the bandwidth of a parallel resonant circuit,
- measure the impedance at frequencies above and below the resonant frequency and observe that it is purely resistive at resonance,
- sketch the output voltage as a function of frequency and explain why the response has a bell-shaped curve when plotted on a semi-logarithmic graph.

EQUIPMENT REQUIRED

☐ Dual trace oscilloscope
☐ Signal generator (sinusoidal function generator)
☐ DMM
  *Note:*Record this equipment in Table 22-1.

COMPONENTS

☐ Resistors: 1-kΩ, 1.5-kΩ (1/4-W carbon, 5% tolerance)
☐ Capacitors: 0.33-µF (10% tolerance)
☐ Inductors: 2.4-mH (iron core, 5% tolerance)

EQUIPMENT USED

Instrument	Manufacturer/Model No.	Serial No.
Oscilloscope		
Signal generator		
DMM		

Table 22-1

TEXT REFERENCE:

Section 21-2   QUALITY FACTOR, *Q*
Section 21-6   PARALLEL RESONANCE

DISCUSSION

Although series resonant networks are used occasionally in electrical and electronic circuits, parallel resonant networks are the most common type used. Figure 22-1 illustrates a simple parallel resonant network, often called an *LC tank circuit*.

The impedance of the tank circuit is relatively low at all frequencies except at the frequency of resonance. At the resonant frequency, the reactance of the capacitor is exactly equal to the reactance of the inductor. The resulting parallel impedance approaches that of an open circuit. Since the inductor will always have some series resistance due to the coil of wire, the actual impedance of the tank circuit will not be infinitely large. In practice, the impedance of the tank circuit will generally have an impedance between 10 kΩ and 100 kΩ. When the tank circuit is connected across a constant current source (usually a transistor), the voltage across the tank circuit will be relatively high at the resonant frequency and very low at all other frequencies.

The resonant frequency of a tank circuit is found to be

$$f_P = \frac{1}{2\pi\sqrt{LC}}\sqrt{1 - \frac{R_{\text{coil}}^2 C}{L}}$$ (22-1)

If R_{coil}^2 is at least ten times smaller than the ratio L/C, then the parallel resonant frequency may be approximated as

$$f_P = \frac{1}{2\pi\sqrt{LC}}$$ (22-2)

The input impedance of a tank circuit at resonance will always be purely resistive, and may be determined by using the *quality factor Q* of the coil as follows:

$$R_P = (Q_{\text{coil}}^2 + 1)R_{\text{coil}}$$ (22-3)

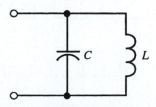

Figure 22-1   Ideal LC tank circuit

If a resistance R_1 is placed in parallel with the tank, the Q of the circuit will be reduced since this resistor absorbs some of the energy from the circuit. Also, if the voltage source has a series resistance R_s, then the Q of the circuit is reduced still further. For the network shown, the quality factor is determined as

$$Q = \frac{R_{eq}}{X_c} \quad \text{where } R_{eq} = R_1 \| R_P \| R_S \qquad (22\text{-}4)$$

Notice that the Q of the circuit is determined by placing R_s in parallel with the other resistors. The reason for this becomes apparent if the volage source and its series resistance are converted into an equivalent current source and parallel resistance.

As in the series resonant circuit, the quality factor may be used to determine the bandwidth of the circuit as

$$BW = \frac{f_P}{Q} \qquad (22\text{-}5)$$

CALCULATIONS

1. Calculate and record the resonant frequency for the circuit of Figure 22-2.

f_P	

2. Prior to starting the lab, obtain an inductor from your lab instructor. Use the DMM ohmmeter to measure the dc resistance of the inductor. Enter the result below.

R_{coil}	

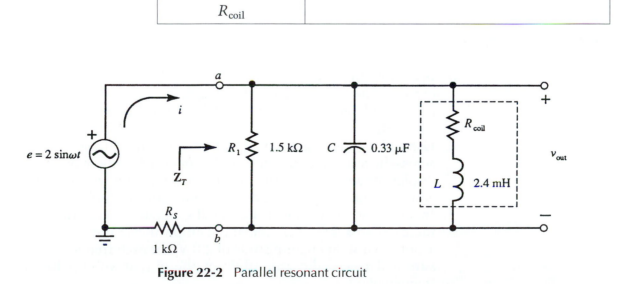

Figure 22-2 Parallel resonant circuit

3. Calculate the equivalent impedance R_P of the LC tank at resonance. Use this value to determine the total impedance $\mathbf{Z}_T$ of the circuit at resonance. (The impedance will be resistive.) Calculate the phasor form of current $\mathbf{I}$ at resonance and determine the output voltage phasor $\mathbf{V}_{\text{out}}$. Enter your results below.

R_P	
$\mathbf{Z}_T$	
$\mathbf{I}$	
$\mathbf{V}_{\text{out}}$	

4. Calculate and record the quality factor Q, bandwidth BW, and half-power frequencies f_1 and f_2 for the circuit.

Q	
BW	
f_1	
f_2	

MEASUREMENTS

5. Assemble the circuit shown in Figure 22-2. Connect Ch1 of the oscilloscope to the output of the signal generator and adjust the output to have an amplitude of 2.0 V_p (4.0 V_{p-p}) at a frequency of $f = 1$ kHz.
6. Use Ch1 as the trigger source and connect Ch2 of the oscilloscope across the resistor R_S. Set the oscilloscope on the difference mode to display the amplitude of the sinusoidal output voltage. Measure the amplitude of voltage V_{out} and record the results in Table 22-2.
7. Increase the frequency of the signal generator to the frequencies indicated in Table 22-2. Adjust the amplitude of the signal generator to maintain an amplitude of 2.0 V_p for each frequency. Measure and record the amplitude of the output voltage for each frequency.

f	V_{out}
1 kHz	
2 kHz	
3 kHz	
4 kHz	
4.5 kHz	
5 kHz	
5.5 kHz	
6 kHz	
6.5 kHz	
7 kHz	
8 kHz	
9 kHz	
10 kHz	

Table 22-2

8. For which frequency f in Table 22-2 is the output voltage a maximum? Adjust the generator to provide an output of 2.0 V_p at this frequency. While observing the oscilloscope, adjust the the generator frequency until voltage V_{out} is at the maximum value. Record the resonant frequency f_P and the corresponding output voltage V_{out} in Table 22-3.

9. Decrease the frequency until the output voltage V_{out} is reduced to 0.707 of the maximum value found in Step 8. Record the lower half-power frequency f_1 and the corresponding output voltage V_{out} in Table 22-3. (Ensure that the output of the signal generator is at 2.0 V_p.)

   Increase the frequency above the resonant frequency until the output voltage V_{out} is again reduced to 0.707 of the maximum value found in Step 8. Record the upper half-power frequency f_2 and the corresponding resistor voltage V_{out} in Table 22-3. (Ensure that the output of the signal generator is at 2.0 V_p.)

f	V_{out}
$f_1 =$	
$f_P =$	
$f_2 =$	

Table 22-3

10. Set the signal generator to the resonant frequency determined in Step 9 and adjust the amplitude for 2.0 V_p. Measure the magnitude and phase angle (with respect to the signal generator) of the voltage across R_S. Calculate the phasor form of current $\mathbf{I}$ at resonance. (You should observe that v_S and e are in phase.) Record your results.

V_S	
θ	
$\mathbf{I}$	

11. Adjust the signal generator for a frequency $f = f_1$ and a voltage of 2.0 V_p. Measure the magnitude and phase angle of the voltage V_S. Calculate the phasor form of current $\mathbf{I}$ at this frequency. Record these values.

V_R	
θ	
$\mathbf{I}$	

12. Adjust the signal generator for a frequency $f = f_2$ and a voltage of 2.0 V_p. Measure the magnitude and phase angle of the voltage V_S. Calculate the phasor form of the current $\mathbf{I}$ at this frequency. Record the values.

V_S	
θ	
$\mathbf{I}$	

CONCLUSIONS

13. Plot the data of Tables 22-2 and 22-3 on the semi-logarithmic scale of Graph 22-1. Connect the points with the best smooth continuous curve.

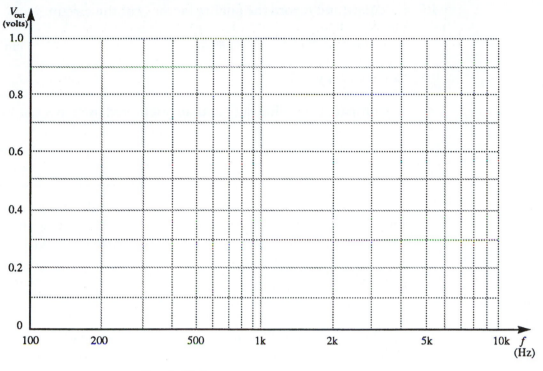

Graph 22-1

14. Compare the measured resonant frequency f_P as found in Step 8 to the theoretical frequency determined in Step 1.

_____

_____

15. Compare the measured half-power frequencies of Step 9 to the theoretical values determined in Step 3.

_____

_____

16. Calculate and record the bandwidth.

$$BW = f_2 - f_1 \qquad (22\text{-}6)$$

BW	

17. Calculate and record the quality factor Q of the circuit.

$$Q = \frac{f_S}{BW}$$

(22-7)

Q	

Compare this value of Q to that calculated in Step 3.

_____

_____

18. Use the data of Step 10 to compare the phase angle of the current with respect to the signal generator at resonance. Based on this result, is the circuit resistive, inductive, or capacitive when $f = f_P$?

_____

_____

19. Use the data of Step 11 to compare the phase angle of the current with respect to the signal generator when $f = f_1$. Based on this result, is the circuit resistive, inductive, or capacitive when $f < f_P$?

_____

_____

20. Use the data of Step 12 to compare the phase angle of the current with respect to the signal generator when $f = f_2$. Based on this result, is the circuit resistive, inductive, or capacitive when $f > f_P$?

_____

_____

FOR FURTHER INVESTIGATION AND DISCUSSION

21. If the resistance of the inductor R_{coil} was higher than the measured value, what would happen to R_P, Q, and BW at resonance?

R_P :

_____

Q :

_____

BW :

_____

RC and RL Low-Pass Filter Circuits

OBJECTIVES

After completing this lab, you will be able to
- develop the transfer function for a low-pass filter circuit,
- determine the cutoff frequency of a low-pass filter circuit,
- sketch the Bode plot of the transfer fuction for a low-pass filter,
- compare the measured voltage gain response of a low pass filter to the theoretical asymptotic response predicted by a Bode plot,
- explain why the voltage gain of a low-pass filter drops at a rate of 20 dB for each decade increase in frequency.

EQUIPMENT REQUIRED

☐ Dual trace oscilloscope
☐ Signal generator (sinusoidal function generator)
☐ DMM
  Note: Record this equipment in Table 23-1.

COMPONENTS

☐ Resistors: 75-Ω, 330-Ω (1/4-W carbon, 5% tolerance)
☐ Capacitors: 0.47-μF (10% tolerance)
☐ Inductors: 2.4-mH (iron core, 5% tolerance)

EQUIPMENT USED

Instrument	Manufacturer/Model No.	Serial No.
Oscilloscope		
Signal generator		

Table 23-1

TEXT REFERENCE

Section 22.3   SIMPLE *RC* AND *RL* TRANSFER FUNCTIONS
Section 22.4   THE LOW-PASS FILTER CIRCUIT

DISCUSSION

Filter circuits are used extensively in electrical and electronic circuits to remove unwanted signals while permitting desired signals to pass from one stage to another. Although there are many types of filter circuits, most filters are *low-pass, high-pass, band-pass,* or *band-reject* filters. As the name implies, *low-pass* filters permit low frequencies to pass from one stage to another. Figure 23-1 shows both *RC* and *RL* low-pass filters.

The frequency at which the low-pass filter begins to attenuate (decrease the amplitude of) a signal is called the *cutoff frequency* or *break frequency* f_C. Specifically, the cutoff frequency is that frequency at which the amplitude of the output voltage is 0.707 of the amplitude of low frequency signals. Since this frequency corresponds to half power, the cutoff frequency is also called the *half-power* or *3-dB down frequency*. (Recall that when the output power is half of the input power, the attenuation of the stage is 3 dB.)

Cutoff frequencies are generally calculated as ω_C in radians per second, since the algebra tends to be fairly straightforward. However, when working with measurements it is easiest to use frequencies expressed as f_C in hertz. The cutoff frequencies for an *RC* low-pass filter are given as follows:

$$\omega_C = \frac{1}{\tau} = \frac{1}{RC} \tag{23-1}$$

and

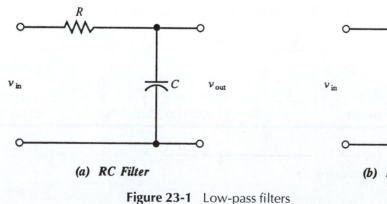

(a) RC Filter

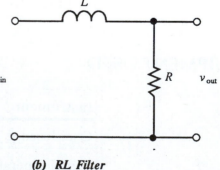

(b) RL Filter

Figure 23-1   Low-pass filters

$$f_C = \frac{\omega_C}{2\pi} = \frac{1}{2\pi RC} \qquad (23\text{-}2)$$

For an *RL* low-pass filter, the cutoff frequencies are

$$\omega_C = \frac{1}{\tau} = \frac{R}{L} \qquad (23\text{-}3)$$

and

$$f_C = \frac{\omega_C}{2\pi} = \frac{R}{2\pi L} \qquad (23\text{-}4)$$

The frequency response of a filter circuit is generally shown on two semi-logarithmic graphs where the abscissa (horizontal axis) for each graph gives the frequency on a logarithmic scale. The ordinate (vertical axis) of one graph shows the voltage gain in decibels, while the ordinate of the other graph shows the phase shift of the output voltage with respect to the applied input voltage. The voltage gain and phase shift for any frequency are determined by finding the *transfer function TF* of the given filter. The transfer function is defined as the ratio of the output voltage phasor to the input voltage phasor.

$$\mathbf{TF} = \frac{\mathbf{V}_{out}}{\mathbf{V}_{in}} \qquad (23\text{-}5)$$

CALCULATIONS

The *RC* Low-Pass Filter

1. Write the transfer function $\mathbf{TF} = \mathbf{V}_{out}/\mathbf{E}$ for the *RC* low-pass filter of Figure 23-2.

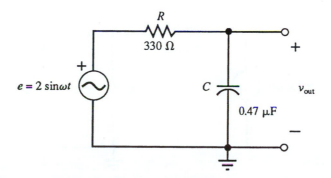

Figure 23-2   *RC* low-pass filter

2. Determine the cutoff frequencies (in radians per second and in hertz) of the circuit in Figure 23-2. Record the values below.

ω_C		rad/s
f_C		Hz

3. Sketch the straight-line approximations of the frequency responses (A_v in dB versus frequency and θ versus frequency) for the *RC* low-pass filter on Graph 23-1.

The *RL* Low-Pass Filter

4. Write the transfer function **TF** = $\mathbf{V}_{out}/\mathbf{E}$ for the *RL* low-pass filter of Figure 23-3.

5. Determine the cutoff frequencies (in radians per second and in hertz) of the circuit in Figure 23-3. Record the values below.

ω_C		rad/s
f_C		Hz

6. Sketch the straight-line approximations of the frequency responses (A_v in dB versus frequency and θ versus frequency) for the *RL* low-pass filter on Graph 23-2.

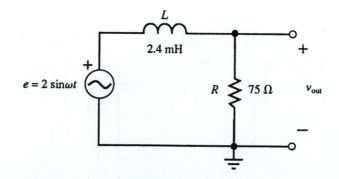

Figure 23-3   *RL* low-pass filter

MEASUREMENTS

The *RC* Low-Pass Filter

7. Assemble the circuit shown in Figure 23-2. Connect Ch1 of the oscilloscope to the output of the signal generator and adjust the output to have an amplitude of 2.0 V_p (4.0 V_{p-p}) at a frequency of $f = 100$ Hz.

8. Use Ch1 as the trigger source and connect Ch2 of the oscilloscope across the capacitor C. Measure the amplitude and phase angle (with respect to e) of the sinusoidal output voltage v_{out}. Enter the results in Table 23-2.

9. Increase the frequency of the signal generator to the frequencies indicated in Table 23-2. If necessary, adjust the amplitude of the signal generator to ensure that the output is maintained at 2.0 V_p (4.0 V_{p-p}). Measure the amplitude and phase shift of v_{out} (with respect to e) for each frequency.

f	V_{out}	
	Amplitude	Phase shift
100 Hz		
200 Hz		
400 Hz		
800 Hz		
1 kHz		
2 kHz		
4 kHz		
8 kHz		
10 kHz		

Table 23-2

10. Determine the cutoff frequency by adjusting the frequency of the generator until the output voltage has an amplitude of $V_{out} = (0.707)(2.0\,V_p) = 1.41\,V_p$. Record the measured cutoff frequency below.

f_C	

The *RL* Low-Pass Filter

11. Construct the circuit shown in Figure 23-3. Connect Ch1 of the oscilloscope to the output of the signal generator and adjust the output to have an amplitude of 2.0 V_p (4.0 V_{p-p}) at a frequency of f = 100 Hz.

12. Use Ch1 as the trigger source and connect Ch2 of the oscilloscope across the resistor R. Measure the amplitude and phase angle (with respect to e) of the sinusoidal output voltage v_{out}. Enter the results in Table 23-3.

13. Increase the frequency of the signal generator to the frequencies indicated in Table 23-3. If necessary, adjust the amplitude of the signal generator to ensure that the output is maintained at 2.0 V_p (4.0 V_{p-p}). Measure the amplitude and phase shift of v_{out} (with respect to e) for each frequency.

f	v_{out}	
	Amplitude	Phase shift
100 Hz		
200 Hz		
400 Hz		
800 Hz		
1 kHz		
2 kHz		
4 kHz		
8 kHz		
10 kHz		

Table 23-3

14. Determine the cutoff frequency by adjusting the frequency of the generator until the output voltage has an amplitude of V_{out} = (0.707)(2.0 V_p) = 1.41 V_p. Record the measured cutoff frequency here.

f_C	

CONCLUSIONS

The *RC* Low-Pass Filter

15. Use your measurements recorded in Table 23-2 to calculate the magnitude of the gain as a ratio of the amplitudes V_{out}/V_{in}. Calculate the voltage gain in decibels as

$$[A_v]_{dB} = 20 \log \frac{V_{out}}{V_{in}} \qquad (23\text{-}6)$$

Record the calculated voltage gain and measured phase shift for each frequency in Table 23-4.

f	$A_v = V_{out}/V_{in}$	$[A_v]_{dB}$	θ
100 Hz			
200 Hz			
400 Hz			
800 Hz			
1 kHz			
2 kHz			
4 kHz			
8 kHz			
10 kHz			

Table 23-4

16. Plot the data of Table 23-4 on the semi-logaritmic scales of Graph 23-1 (page 181). Connect the points with the best smooth continuous curve. You should find that the actual response is closely predicted by the straight-line approximations of the Bode plot.
17. Compare the measured cutoff frequency f_C of Step 10 to the theoretical value predicted by the transfer function in Step 2.

The *RL* Low-Pass Filter

18. Use your measurements recorded in Table 23-3 to calculate the magnitude of the gain as a ratio of the amplitudes V_{out}/V_{in}. Determine the voltage gain in decibels. Record the calculated voltage gain and measured phase shift for each frequency in Table 23-5.

f	$A_v = V_{out}/V_{in}$	$[A_v]_{dB}$	θ
100 Hz			
200 Hz			
400 Hz			
800 Hz			
1 kHz			
2 kHz			
4 kHz			
8 kHz			
10 kHz			

Table 23-5

19. Plot the data of Table 23-5 on the semi-logarithmic scales of Graph 23-2 (page 176). Connect the points with the best smooth continuous curve. You should find that the actual response is closely predicted by the straight-line approximations of the Bode plot.

20. Compare the measured cutoff frequency f_C of Step 14 to the theoretical value predicted by the transfer function in Step 5.

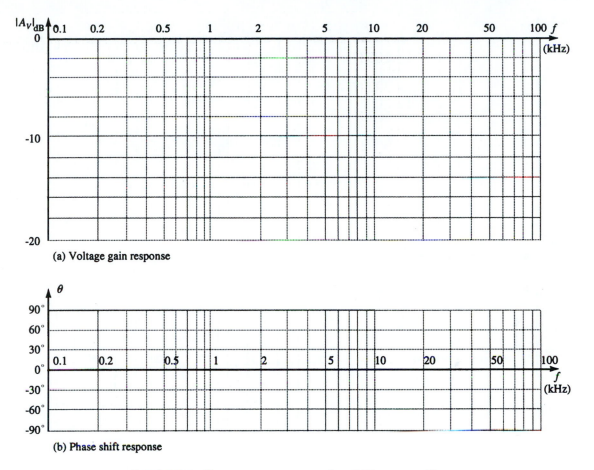

(a) Voltage gain response

(b) Phase shift response

Graph 23-1   Frequency response of an *RC* low-pass filter

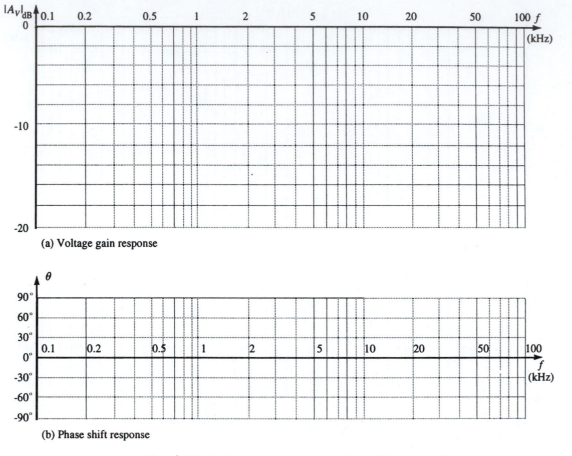

(a) Voltage gain response

(b) Phase shift response

Graph 23-2   Frequency response of an *RL* low-pass filter

FOR FURTHER INVESTIGATION AND DISCUSSION

21. Use MultiSIM or PSpice to simulate the circuit of Figure 23-2. Obtain the frequency response curves for this circuit and compare the results to the actual measurements.
22. Repeat Step 21 for the circuit of Figure 23-3.

NAME _____

DATE _____

CLASS _____

<table>
<tr><td>LAB
24</td><td></td></tr>
</table>

RC and *RL* High-Pass Filter Circuits

OBJECTIVES

After completing this lab, you will be able to
- develop the transfer function for a high-pass filter circuit,
- determine the cutoff frequency of a high-pass filter circuit,
- sketch the Bode plot of the transfer fuction for a high-pass filter,
- compare the measured voltage gain response of a high-pass filter to the theoretical asymptotic response predicted by a Bode plot,
- explain why the voltage gain of a high-pass filter increases at a rate of 20 dB/decade below the cutoff frequency.

EQUIPMENT REQUIRED

☐ Dual trace oscilloscope
☐ Signal generator (sinusoidal function generator)
☐ DMM
Note: Record this equipment in Table 24-1.

COMPONENTS

☐ Resistors: 75-Ω, 330-Ω (1/4-W carbon, 5% tolerance)
☐ Capacitors: 0.47-μF (10% tolerance)
☐ Inductors: 2.4-mH (iron core, 5% tolerance)

EQUIPMENT USED

Instrument	Manufacturer/Model No.	Serial No.
Oscilloscope		
Signal generator		

Table 24-1

TEXT REFERENCE

Section 22.5   THE HIGH-PASS FILTER CIRCUIT

DISCUSSION

As the name implies, the *high-pass filter* permits high frequencies to pass from the input through to the output of the filter. Due to the abundance of electric motors and fluorescent lights, *60-Hz noise* is the most prevalent unwanted signal around us. Although many applications exist, one of the most common uses for the high-pass filter is to prevent 60-Hz noise from entering a sensitive electrical or electronic system. Figure 24-1 shows both *RC* and *RL* high-pass filters.

As in the low-pass filter circuit, the *cutoff frequency* f_C is that frequency at which the amplitude of the output voltage is 0.707 of the maximum amplitude. However, in the case of high-pass filters, the maximum amplitude occurs for high frequencies rather than for low frequencies. Since this frequency corresponds to half power, the cutoff frequency is also called the *half-power* or *3-dB down frequency*.

The cutoff frequencies for an *RC* high-pass filter are identical to the values for the low-pass *RC* filter and are given as follows:

$$\omega_C = \frac{1}{\tau} = \frac{1}{RC} \tag{24-1}$$

and

$$f_C = \frac{\omega_C}{2\pi} = \frac{1}{2\pi RC} \tag{24-2}$$

Similarly, for an *RL* high-pass filter, the cutoff frequencies are

$$\omega_C = \frac{1}{\tau} = \frac{R}{L} \tag{24-3}$$

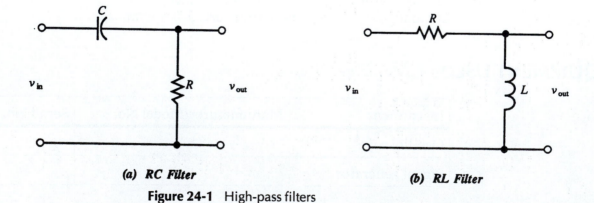

(a) **RC Filter**   *(b)* **RL Filter**

Figure 24-1   High-pass filters

and

$$f_C = \frac{\omega_C}{2\pi} = \frac{R}{2\pi L} \qquad (24\text{-}4)$$

As in the low-pass filter, the frequency response of a high-pass filter circuit is generally shown on two semi-logarithmic graphs where the abscissa (horizontal axis) for each graph gives the frequency on a logarithmic scale. The ordinate (vertical axis) of one graph shows the voltage gain in decibels, while the ordinate of the other graph shows the phase shift of the output voltage with respect to the applied input voltage. The voltage gain and phase shift for any frequency are determined by finding the *transfer function TF* of the given filter. Recall that the transfer function is defined as the ratio of the output voltage phasor to the input voltage phasor.

$$\mathbf{TF} = \frac{\mathbf{V}_{\text{out}}}{\mathbf{V}_{\text{in}}} \qquad (24\text{-}5)$$

CALCULATIONS

The *RC* High-Pass Filter

1. Write the transfer function $\mathbf{TF} = \mathbf{V}_{\text{out}}/\mathbf{E}$ for the *RC* high-pass filter of Figure 24-2.

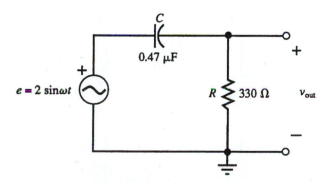

Figure 24-2   *RC* high-pass filter

2. Determine the cutoff frequencies (in radians per second and in hertz) of the circuit in Figure 24-2. Record the values below.

ω_C		rad/s
f_C		Hz

3. Sketch the straight-line approximations of the frequency responses (A_v in dB versus frequency and θ versus frequency) for the *RC* high-pass filter on Graph 24-1 (page 190).

The *RL* High-Pass Filter

4. Write the transfer function **TF** = $\mathbf{V}_{out}/\mathbf{E}$ for the *RL* high-pass filter of Figure 24-3.

5. Determine the cutoff frequencies (in radians per second and in hertz) of the circuit in Figure 24-3. Record the values below.

ω_C		rad/s
f_C		Hz

6. Sketch the straight-line approximations of the frequency responses (A_v in dB versus frequency and θ versus frequency) for the *RL* high-pass filter on Graph 24-2 (page 191).

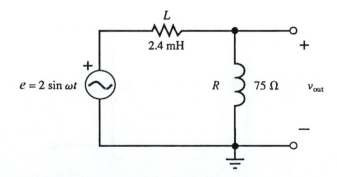

Figure 24-3 *RL* high-pass filter

MEASUREMENTS

The *RC* High-Pass Filter

7. Assemble the circuit shown in Figure 24-2. Connect Ch1 of the oscilloscope to the output of the signal generator and adjust the output to have an amplitude of 2.0 V_p (4.0 V_{p-p}) at a frequency of f = 100 Hz.

8. Use Ch1 as the trigger source and connect Ch2 of the oscilloscope across the resistor R. Measure the amplitude and phase angle (with respect to e) of the sinusoidal output voltage v_{out}. Enter the results in Table 24-2.

9. Increase the frequency of the signal generator to the frequencies indicated in Table 24-2. If necessary, adjust the amplitude of the signal generator to ensure that the output is maintained at 2.0 V_p (4.0 V_{p-p}). Measure the amplitude and phase shift of v_{out} (with respect to e) for each frequency.

f	v_{out}	
	Amplitude	Phase shift
100 Hz		
200 Hz		
400 Hz		
800 Hz		
1 kHz		
2 kHz		
4 kHz		
8 kHz		
10 kHz		

Table 24-2

10. Determine the cutoff frequency by adjusting the frequency of the generator until the output voltage has an amplitude of V_{out} = (0.707)(2.0 V_p) = 1.41 V_p. Record the measured cutoff frequency below.

f_C	

The *RL* High-Pass Filter

11. Construct the circuit shown in Figure 24-3. Connect Ch1 of the oscilloscope to the output of the signal generator and adjust the output to have an amplitude of 2.0 V_p (4.0 V_{p-p}) at a frequency of $f = 100$ Hz.

12. Use Ch1 as the trigger source and connect Ch2 of the oscilloscope across the inductor L. Measure the amplitude and phase angle (with respect to e) of the sinusoidal output voltage v_{out}. Enter the results in Table 24-3.

13. Increase the frequency of the signal generator to the frequencies indicated in Table 24-3. If necessary, adjust the amplitude of the signal generator to ensure that the output is maintained at 2.0 V_p (4.0 V_{p-p}). Measure the amplitude and phase shift of v_{out} (with respect to e) for each frequency.

f	V_{out}	
	Amplitude	Phase shift
100 Hz		
200 Hz		
400 Hz		
800 Hz		
1 kHz		
2 kHz		
4 kHz		
8 kHz		
10 kHz		

Table 24-3

14. Determine the cutoff frequency by adjusting the frequency of the generator until the output voltage has an amplitude of $V_{out} = (0.707)(2.0\,V_p) = 1.41\,V_p$. Record the measured cutoff frequency below.

f_C	

CONCLUSIONS

The *RC* High-Pass Filter

15. Use the measurements recorded in Table 24-2 to calculate the magnitude of the gain as a ratio of the amplitudes V_{out}/V_{in}. Calculate the voltage gain in decibels as

$$[A_v]_{dB} = 20 \log \frac{V_{out}}{V_{in}}$$
(24-6)

Record the calculated voltage gain and measured phase shift for each frequency in Table 24-4.

f	$A_v = V_{out}/V_{in}$	$[A_v]_{dB}$	θ
100 Hz			
200 Hz			
400 Hz			
800 Hz			
1 kHz			
2 kHz			
4 kHz			
8 kHz			
10 kHz			

Table 24-4

16. Plot the data of Table 24-4 on the semi-logaritmic scales of Graph 24-1 (next page). Connect the points with the best smooth continuous curve. You should find that the actual response is closely approximated by the Bode plot.

17. Compare the measured cutoff frequency f_C of Step 10 to the theoretical value predicted by the transfer function in Step 2.

_____

_____

The *RL* High-Pass Filter

18. Use the measurements recorded in Table 24-3 to calculate the magnitude of the gain as a ratio of the amplitudes V_{out}/V_{in}. Determine the voltage gain in decibels. Record the calculated voltage gain and measured phase shift for each frequency in Table 24-5.

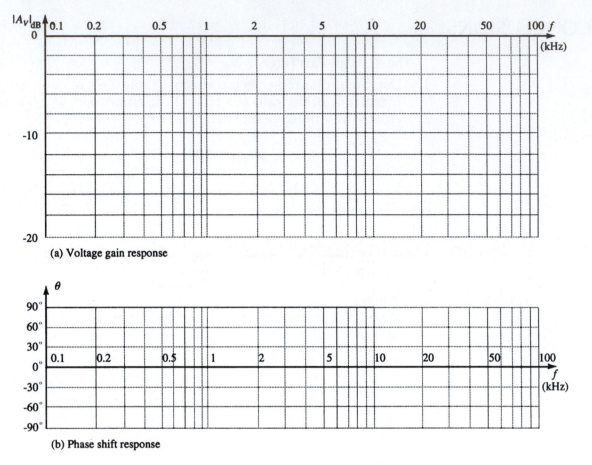

(a) Voltage gain response

(b) Phase shift response

Graph 24-1   Frequency response of an *RC* high-pass filter

f	$A_v = V_{out}/V_{in}$	$[A_v]_{dB}$	θ
100 Hz			
200 Hz			
400 Hz			
800 Hz			
1 kHz			
2 kHz			
4 kHz			
8 kHz			
10 kHz			

Table 24-5

19. Plot the data of Table 24-5 on the semi-logaritmic scales of Graph 24-2. Connect the points with the best smooth continuous curve. You should find that the actual response is closely approximated by the Bode plot.
20. Compare the measured cutoff frequency f_C of Step 14 to the theoretical value predicted by the transfer function in Step 5.

_____

_____

FOR FURTHER INVESTIGATION AND DISCUSSION

21. Use MultiSIM or PSpice to simulate the circuit of Figure 24-2. Obtain the frequency response curves for this circuit and compare the results to the actual measurements.
22. Repeat Step 21 for the circuit of Figure 24-3.

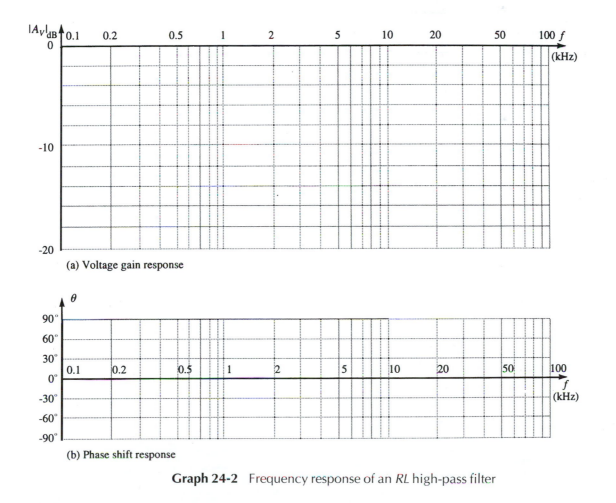

(a) Voltage gain response

(b) Phase shift response

Graph 24-2   Frequency response of an *RL* high-pass filter

LAB 25

Bandpass Filter

OBJECTIVES

After completing this lab, you will be able to
- calculate the cutoff frequencies of a bandpass filter by examining the individual low-pass and high-pass stages of a filter,
- derive the transfer function for each stage of a bandpass filter,
- sketch the Bode plot of a bandpass filter from the transfer function of the individual stages,
- explain why the slopes of the voltage gain response is 20 dB/decade on each side of the cutoff frequencies.

EQUIPMENT REQUIRED

☐ Dual trace oscilloscope
☐ Signal generator (sinusoidal function generator)
☐ DMM
   Note: Record this equipment in Table 25-1.

COMPONENTS

☐ Resistors: 330-Ω (2) (1/4-W carbon, 5% tolerance)
☐ Capacitors: 0.047-μF, 0.47-μF (10% tolerance)

EQUIPMENT USED

Instrument	Manufacturer/Model No.	Serial No.
Oscilloscope		
Signal generator		

Table 25-1

TEXT REFERENCE

Section 22.6   BANDPASS FILTER

DISCUSSION

Bandpass filters permit a range of frequencies to pass from one stage to another. Many different types of bandpass filters are used throughout electronics. For instance, the typical television receiver uses several stages of filtering to achieve a bandwidth of 6 MHz, while an AM receiver has circuitry which restricts the bandwidth to only 10 kHz. Although bandpass filters can be very elaborate, depending on the application, they can also be constructed simply by combining a low-pass filter and a high-pass filter as shown in Figure 25-1. Since expense and other design difficulties make *RL* filters impractical, *RC* filters are used almost exclusively.

The bandpass filter of Figure 25-1 has two cutoff frequencies as determined by the cutoff frequencies of the individual stages. The frequency below which the high-pass filter attenuates the signal is called the *lower cutoff frequency f_1*. The frequency at which the low-pass filter begins to attenuate is called the *upper cutoff frequency f_2*. As expected, the difference between the two frequencies is the *bandwidth BW*. In order for the circuit to operate predictably, the cutoff frequencies should be separated by at least one *decade* and the first stage must be the high-pass circuit.

The lower cutoff frequency ω_1 in radians per second is found as

$$\omega_1 = \frac{1}{\tau_1} = \frac{1}{R_1 C_1} \tag{25-1}$$

with a corresponding frequency f_1 in hertz,

$$f_1 = \frac{\omega_1}{2\pi} = \frac{1}{2\pi R_1 C_1} \tag{25-2}$$

The upper cutoff frequency ω_2 in radians per second is

$$\omega_2 = \frac{1}{\tau_2} = \frac{1}{R_2 C_2} \tag{25-3}$$

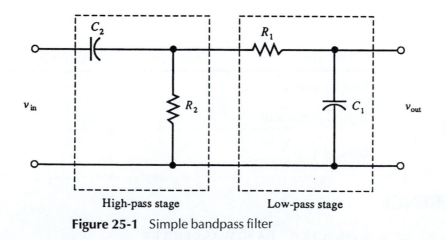

Figure 25-1   Simple bandpass filter

with a corresponding frequency f_2 in hertz,

$$f_2 = \frac{\omega_2}{2\pi} = \frac{1}{2\pi R_2 C_2}$$ (25–4)

The bandwidth of the filter is determined as

$$BW = f_2 - f_1$$ (25-5)

CALCULATIONS

1. Calculate and record the cutoff frequency (in radians per second and in hertz) for the high-pass stage in the circuit of Figure 25-2.

ω_1		rad/s
f_1		Hz

2. Calculate and record the cutoff frequency (in radians per second and in hertz) for the low-pass stage in the circuit of Figure 25-2.

ω_2		rad/s
f_2		Hz

3. Calculate and record the bandwidth of the filter circuit of Figure 25-2.

BW		Hz

4. From the calculations of Steps 1 and 2 sketch the Bode plots for the bandpass filter of Figure 25-2. Use the semilogarithmic scales of Graph 25-1 (page 198).

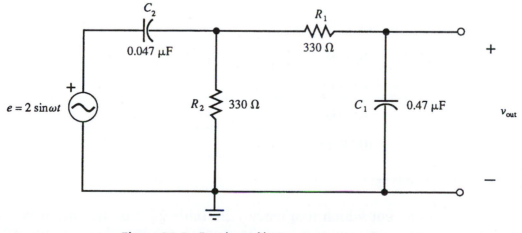

Figure 25-2 Bandpass filter

MEASUREMENTS

5. Assemble the circuit shown in Figure 25-2. Connect Ch1 of the oscilloscope to the output of the signal generator and adjust the output to have an amplitude of 2.0 V_p (4.0 V_{p-p}) at a frequency of $f = 100$ Hz.
6. Use Ch1 as the trigger source and connect Ch2 of the oscilloscope across the output of the circuit. Measure the amplitude and phase angle of the output voltage v_{out} (with respect to e). Enter the results in Table 25-2.
7. Increase the frequency of the signal generator to the frequencies indicated in Table 25-2. Due to loading effects of the circuit, it will be necessary to readjust the amplitude of the signal generator at each frequency to ensure that the output is maintained at 2.0 V_p (4.0 V_{p-p}). Measure and record the amplitude and phase shift of the output voltage v_{out} (with respect to e) for each frequency.

f	v_{out}	
	Amplitude	Phase shift
100 Hz		
200 Hz		
400 Hz		
800 Hz		
1 kHz		
2 kHz		
4 kHz		
8 kHz		
10 kHz		
20 kHz		
40 kHz		
80 kHz		
100 kHz		

Table 25-2

8. For which frequency f in Table 25-2 is the output voltage a maximum? Adjust the generator to provide an output of 2.0 V_p (4.0 V_{p-p}) at this frequency. While observing the oscilloscope, ad-

just the the generator frequency until voltage V_{out} is at the maximum value. Measure and record the center frequency f_0 and the corresponding amplitude V_{out}.

f_0	

9. Decrease the frequency until the output voltage V_{out} is reduced to 0.707 of the maximum value found in Step 8. Record the lower half-power frequency f_1 and the corresponding output voltage V_{out}. (Ensure that the output of the signal generator is at 2.0 V_p.)

f_1	

10. Increase the frequency above the center frequency until the output voltage V_{out} is again reduced to 0.707 of the maximum value found in Step 8. Record the upper half-power frequency f_2 and the corresponding resistor voltage V_{out}. (Ensure that the output of the signal generator is at 2.0 V_p.)

f_2	

CONCLUSIONS

11. For each voltage measurement in Table 25-2, determine the voltage gain as both a ratio of amplitudes $A_v = v_{out}/E$ and in decibels as

$$[A_v]_{dB} = 20 \log \frac{V_{out}}{E} \qquad (25\text{-}6)$$

   Record the calculated voltage gain and measured phase shift for each frequency in Table 25-3.
12. Plot the data of Table 25-3 on the semi-logaritmic scale of Graph 25-2. Connect the points with the best smooth continuous curve.
13. Compare the measured cutoff frequencies of Step 9 and 10 to the theoretical frequencies determined in Steps 1 and 2.

_____

_____

14. How do the actual voltage gain and phase shift responses of the output voltage compare to the predicted responses?

_____

_____

_____

f	$A_v = Vout/E$	$[A_v]_{dB}$	θ
100 Hz			
200 Hz			
400 Hz			
800 Hz			
1 kHz			
2 kHz			
4 kHz			
8 kHz			
10 kHz			
20 kHz			
40 kHz			
80 kHz			
100 kHz			

Table 25-3

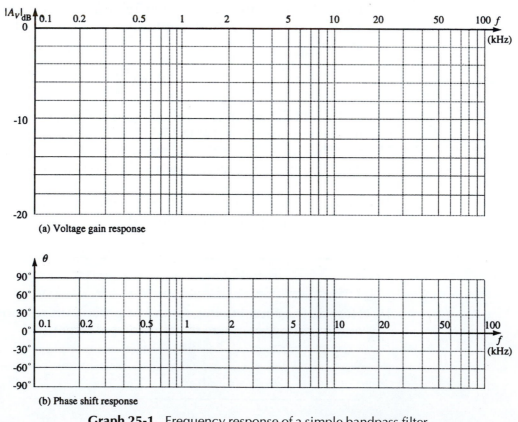

(a) Voltage gain response

(b) Phase shift response

Graph 25-1   Frequency response of a simple bandpass filter

NAME _____

DATE _____

CLASS _____

Voltage, Current and Power in Balanced Three-Phase Systems

OBJECTIVES

After completing this lab, you will be able to
- verify line and phase voltage relationships for a Y-load,
- verify line and phase current relationships for a Δ-load,
- verify the single-phase equivalent method of analysis.
- measure power in a three-phase system

EQUIPMENT REQUIRED

☐ ac ammeters (DMMs with 2-A ranges are adequate)
☐ DMMs
☐ Single Phase wattmeter (2 required)

POWER SUPPLY

☐ Three-phase 120/208 V (60 Hz), preferably variable such as by means of a 3-phase autotransformer

COMPONENTS

☐ Resistors: 100-Ω, 200-W (3), 250-Ω, 200-W (3)
☐ Capacitors: 10 μF, non-electrolytic, rated for operation at 120 VAC (3 required)

EQUIPMENT USED

Instrument	Manufacturer/Model No.	Serial No.
Single-phase wattmeters		
ac Ammeter or DMM		
DMMs		

Table 26-1

TEXT REFERENCE

Section 24.1   THREE-PHASE VOLTAGE GENERATION
Section 24.3   BASIC THREE-PHASE RELATIONSHIPS
Section 24.5   POWER IN BALANCED SYSTEMS
Section 24.6   MEASURING POWER IN THREE-PHASE CIRCUITS

DISCUSSION

Y-Loads. For a Y-Load, Figure 26-1, line-to-line voltages are $\sqrt{3}$ times line-to-neutral voltages and each line-to-line voltage leads its corresponding line-to-neutral voltage by 30°. Thus,

$$\mathbf{V}_{ab} = \sqrt{3}\ \mathbf{V}_{an}\angle 30° \qquad (26\text{-}1)$$

Current in the neutral of a balanced system (if a neutral line is present), is zero.

Δ-Loads. The magnitude of line current for a balanced Δ-Load, Figure 26-2, is $\sqrt{3}$ times the magnitude of the phase current and each line current lags its corresponding phase current by 30°. Thus,

$$\mathbf{I}_a = \sqrt{3}\ \mathbf{I}_{ab}\angle -30° \qquad (26\text{-}2)$$

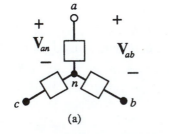

(a)

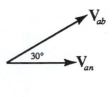

(b)

(a) Circuit

(b) Voltage relationships

Figure 26-1   A balanced Y-load

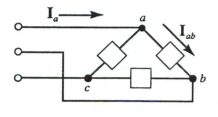

(a) Circuit   *(b) Current relationships*

Figure 26-2   A balanced Δ-load

Equivalent Y and Δ-Loads. For purposes of analysis, a balanced Δ-load may be replaced by an equivalent balanced Y-load with

$$\mathbf{Z}_Y = \mathbf{Z}_\Delta/3 \tag{26-3}$$

Single-Phase Equivalent. Since all three phases of a balanced system are identical, any one phase can be taken to represent the behaviour of all. This permits you to reduce a circuit to its *single-phase equivalent*.

Power in Three-Phase Systems. Power in a balanced three-phase system is three times the power to one phase. Power to one phase is given by

$$P_\phi = V_\phi I_\phi \cos \theta_\phi \tag{26-4}$$

where V_ϕ and I_ϕ are the magnitudes of the phase voltage and current and θ_ϕ, the angle between them, is the angle of the phase load imped-ance. This formula applies to both Y and Δ loads. Total power is three times this.

An alternate power formula based on line voltages and currents is

$$P_T = \sqrt{3}\ V_L I_L \cos \theta_\phi \tag{26-5}$$

where V_L and I_L are the magnitudes of the line-to-line voltages and line currents respectively. This formula applies to both Y and Δ loads. Note that the angle in this formula is the angle of the load im-pedance—it is not the angle between $\mathbf{V}_L$ and $\mathbf{I}_L$.

Measuring Power in Three-Phase Systems. For a 3-wire load, you need only two wattmeters, while for a 4-wire load, you need three wattmeters, Figure 26-3. Note however, the use of individual watt-meters as illustrated here is giving way to the use of integrated, multifunction meters (recall Figure 17-23) that incorporate power/energy measurement for all 3 phases in a single package.

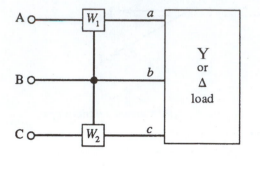

(a) Two-wattmeter method

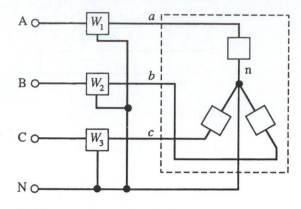

(b) Three-wattmeter method

Figure 26-3 Measuring power in three-phase systems.

MEASUREMENTS

General Safety Notes

1. With power off, assemble your circuit and have your lab partner double check it.

2. Have your instructor check the circuit before you energize it. If you are using a Powerstat or other variable ac source, gradually increase voltage from zero, watching the meters for signs of trouble.

3. Turn power off before changing your circuit for the next test.

Check with your instructor for specific safety instructions.

PART A: Voltages and Currents for a Y-Load

1. With power off, set up the circuit of Figure 26-4. Use DMMs with a current range of 2 A or more to measure current. Observe all saftey precautions—see box.
   a. Set E_{AN} to 120 V (or as close as you can get it) using a DMM, then measure all line-to-neutral and line-to-line voltages at the load. Measure all currents (line and neutral). Record in Table 26.2. Do your measurements confirm the magnitude relationship of Eq. 26-1?

V_{an}	V_{bn}	V_{cn}	V_{ab}	V_{bc}	V_{ca}	I_a	I_b	I_c	I_n

Table 26-2

   b. Turn power off and remove the neutral conductor from between n and N. Measure the line currents again. Are they the same as in Test 1(a)?
   c. Using the measured source voltage and load resistance, compute the magnitude of the line current and compare to the measured value.

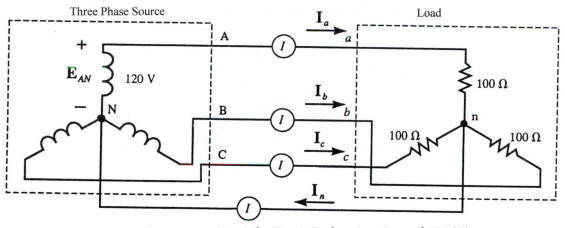

Figure 26-4 Circuit for Test 1. Each resistor is rated 200-W.

PART B: Power to a Y-Load

2. With power off, add wattmeters as in Figure 26-5. Adjust E_{AN} to the same value you used in Part A. (Check currents and note if any have changed. They shouldn't have.)

    $W_1 =$ _____ $W_2 =$ _____

3. Using circuit analysis, determine P_T. Now sum the wattmeter readings. How well do results agree?
4. Using circuit analysis, determine what W_1 should read. Compare it to the actual reading. Repeat for W_2.

PART C: Power to a Y-Load (Continued)

5. De-energize the circuit and add capacitors as in Figure 26-6. Set the voltage as in Part A. Measure line current and power and record.

    $I =$ _____ $W_1 =$ _____ $W_2 =$ _____

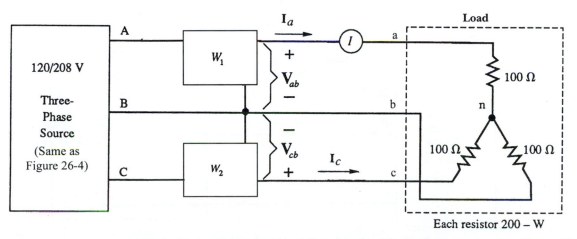

Each resistor 200 – W

Figure 26-5 Circuit for Test 2 (For details of the 3-phase source, see Figure 26-4

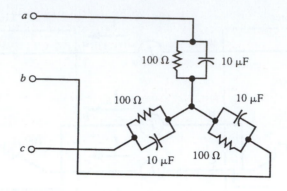

Figure 26-6   Load for Test 5

   a Although the wattmeter readings change, total power should remain unchanged. Why? Verify by using circuit analysis techniques to compute P_T.

   b. Using circuit analysis techniques, compute the reading of W_1. Compare to the measured value. Repeat for W_2.

PART D: Currents for a Δ-Load

6. Assemble the circuit of Figure 26-7, using 250-Ω resistors. Measure line current I_a and phase current I_{ab}. Do your measurements confirm the magnitude relationship of Eq. 26-2?

$I_a = \underline{\hspace{3cm}}$ $I_{ab} = \underline{\hspace{3cm}}$

Part E: Power to a Δ-Load

7. De-energize the circuit and add wattmeters as in Figure 26-3(a). Set the voltage as in Test A. Measure power and record.

$W_1 = \underline{\hspace{3cm}}$ $W_2 = \underline{\hspace{3cm}}$

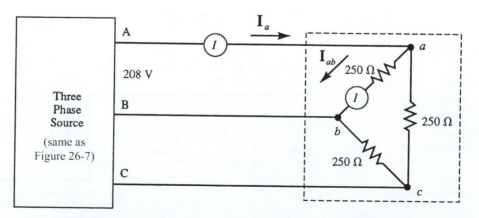

Figure 26-7   Circuit for Test 6. For three phase source, see Figure 26-4.

8. Using circuit analysis techniques, analyze the circuit of Figure 26-7, computing line current, total power and the reading of each wattmeter. Compare your computed results to the measured results. How well do they agree?

PART F: The Single Phase Equivalent

9. Consider the three-phase loads of Figure 26-8(a). (Do not assemble this circuit.) Each resistor of the Y is 250 Ω and each resistor of the Δ is 300 Ω. Source voltage is 120-V, line-to-neutral.
   a. Convert the Δ load to a Y-equivalent then convert the circuit to its single-phase equivalent. Sketch as Figure 26-8(b).
   b. Assemble the single-phase equivalent. Set the source voltage as in Test A and measure line current.

   $E_{AN} = $ _____ $I_A = $ _____

   c. Analyze the single-phase equivalent and compare calculated current I_A to that measured in (b).

PROBLEMS

10. Assume each resistor of the Δ-load of Figure 26-8(a) is replaced with a 10-F capacitor.
    a. Convert the circuit to its single-phase equivalent and sketch.

    b. Calculate the magnitude of the line current I_A.

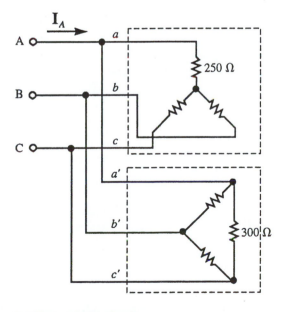

(a) Three-phase circuit   *(b) Single phase equivalent*

Figure 26-8   Circuit for Test 9

    c. Calculate the magnitude of the phase current in the Δ-load.
    d. Calculate total power to the combined load.

COMPUTER ANALYSIS

11. Using MultiSIM or PSpice, verify the currents computed in Question 10.

FOR FURTHER INVESTIGATION AND DISCUSSION

MultiSIM Users
The version of MultiSIM current at the time of writing does not permit you to do this full investigation, as it is only able to plot voltages with respect to ground. Because of this limitation, the only part of this exercise that you can do is Part a, phase voltages.

Using MultiSIM or PSpice, set up the three-phase source of Figure 26-4 and investigate three-phase voltage waveforms. (Since you need time varying waveforms here, use VSIN as your source if you are using PSpice and select Transient Analysis.) Use the following to guide your analysis.

a. Plot waveforms for phase voltages v_{AN}, v_{BN} and v_{CN}.
b. Repeat for line voltages v_{AB}, v_{BC} and v_{CA}.
c. Plot v_{AN} and v_{AB} on the same graph; then, using the cursor, measure values and verify the magnitude and phase relationships of Eq. 26-1.

Write a short report confirming how your study verifies the results discussed in the text.

LAB 27

The Iron-Core Transformer

OBJECTIVE

After completing this lab, you will be able to
- verify the turns ratio and phase relationships for a transformer,
- verify the concept of reflected impedance,
- determine the frequency response of an audio transformer,
- measure the regulation of a power transformer.

Safety Note
In parts of this lab, you will be working with 120 VAC. This is a dangerous voltage and you must be aware of and observe safety precautions. Familiarize yourself with your lab's safety practices and use extreme caution at all times. Always ensure that power is off when you are assembling, changing or otherwise wotking on circuits with dangerous voltages.

EQUIPMENT REQUIRED

☐ Oscilloscope
☐ DMM
☐ Signal or function generator

COMPONENTS

☐ Resistors: 10-Ω, 100-Ω (each 1/4 W), 10-Ω, 25-W
☐ Audio transformer (Hammond 145F or equivalent)
☐ 120/12.6-V filament transformer (Hammond 167L12 or equivalent)

EQUIPMENT USED

Instrument	Manufacturer/Model No.	Serial No.
Oscilloscope		
DMM		
Signal or function generator		

Table 27-1

TEXT REFERENCE

Section 23.2   THE IRON-CORE TRANSFORMER: THE IDEAL
                MODEL
Section 23.3   REFLECTED IMPEDANCE
Section 23.8   VOLTAGE AND FREQUENCY EFFECTS

DISCUSSION

We look first at the ideal transformer. An ideal transformer, Figure
27-1(a), is one that is characterized by its turns ratio

$$a = N_p/N_s \tag{27-1}$$

where N_p and N_s are its primary and secondary turns respectively.
Voltage and current ratios are given by

$$\mathbf{V}_p/\mathbf{V}_s = V_p/V_s = a \tag{27-2}$$

$$\mathbf{I}_p/\mathbf{I}_s = I_p/I_s = 1/a \tag{27-3}$$

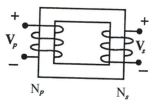

(a) Pictorial

Reflected Impedance. Load impedance $\mathbf{Z}_L$, Figure 27-1(c), is reflected
into the primary as

$$\mathbf{Z_p} = a^2\,\mathbf{Z_L} \tag{27-4}$$

All impedances in the secondary (whether part of the load or not) are
reflected in this fashion.

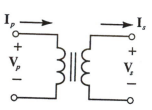

(b) Schematic

Phasing. Depending on the relative direction of windings, the secondary voltage is either in-phase or 180° out-of-phase with respect to the
primary—see Practical Note 1.

Real Transformers. Real transformers differ from the ideal in that
they have primary and secondary winding resistances, leakage flux,
core losses and a few other non-ideal characteristics as described in
Sections 23.6 to 23.8 of the text. In this lab, you will observe some of
these—you will look in particular at the regulation of a power transformer and the bandwidth limitations of an audio transformer.

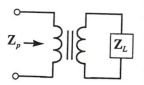

(c) Reflected impedance

Figure 27-1   The ideal
transformer

Voltage Ratios. The voltage ratio of an iron-core transformer under
no-load is the same as its turns ratio. However (because of internal
voltage drops), the ratio under load is different—see Note 2.

MEASUREMENTS

PART A: Turns Ratio

1. Consider Figure 27-2. Observing suitable safety precautions, apply
    full rated voltage to the transformer.

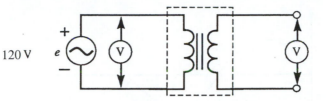

Figure 27-2   Circuit for Test 1. Use a filament transformer (or equivalent) with a 120 V primary and a low voltage (about 20 V or less) secondary.

Practical Notes

1. Small transformers usually have color coded leads to identify corresponding (dotted) terminals while power transformers often use letter designations H_1, X_1, etc.—check the data sheet or nameplate of your transformer for details.

2. Some data sheets list the ratio of primary voltage to secondary voltage under load, some under no load and some list both. For example, the Hammond 167L12 suggested for this lab has a no-load specification of 115V/13.8V and a full-load specification of 115V/12.6V.

a. Measure primary and secondary voltages.

$$V_p = \underline{\hspace{2cm}}; V_s = \underline{\hspace{2cm}}$$

Determine the turns ratio using Equation 27-2.

$$a_{(measured)} = \underline{\hspace{2.5cm}}$$

b. Determine the turns ratio from its no-load data sheet and compare it to that of Test 1(a). $a_{(Name-plate)} = \underline{\hspace{2.5cm}}$

PART B: Phase Relationships

2. Using the audio transformer, assemble the circuit of Figure 27-3. Set the source to a 1-kHz sine wave with an amplitude of 8 V (i.e., 16 V_{p-p}). Mark the transformer terminals for reference in determining phase relationships.

a. Measure the magnitude and phase of v_{cd} relative to the primary voltage and sketch in Figure 27-4. Now reverse the secondary scope leads and sketch v_{dc}. Based on these observations, add the missing dot to the transformer of Figure 27-3. Explain how you determined the position of this dot.

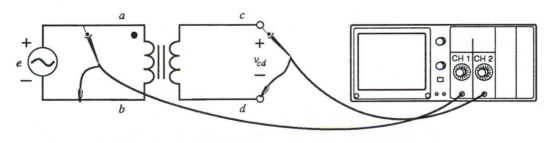

Figure 27-3   Circuit for Test 2 (use the audio transformer)

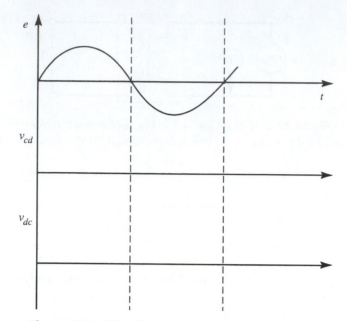

Figure 27-4   Waveforms for Test 2(a)

b. Based on the measurements of Test 2(a), what is the turns ratio for this transformer? a = _____

PART C: Reflected Impedance

3. Measure the 10-Ω sensing resistor and the 100-Ω load resistor, then assemble the circuit of Figure 27-5, using the audio transformer. Connect probes as shown.

R_{sen} = _____ R_{L} = _____

a. Using the differential mode of the oscilloscope, set the input voltage V_{p} of the transformer to 16 V peak-to-peak, 1-kHz.

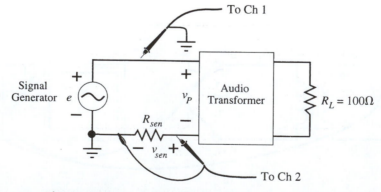

Figure 27-5   Circuit for Test 3

b. Select Ch2 and measure the voltage across the sensing resistor. From this, calculate the primary current.

$V_{\text{sen}} =$ _____ $I_{\text{p}} =$ _____

c. Using the results of Test 3(a) and (b), determine the magnitude of the input impedance to the transformer.

$Z_{\text{in(measured)}} =$ _____

d. For an ideal transformer, input impedance is $Z_{\text{p}} = a^2 Z_{\text{L}}$. However, real transformers have winding resistance and leakage reactance. Neglecting leakage reactance, we get the circuit of Figure 27-6. The winding resistance can now be reflected into the primary along with the load resistance. Adding to this, the primary winding resistance, we get (approximately) an input impedance Z_{in} of

$$Z_{\text{in}} = R_{\text{p}} + a^2(R_{\text{s}} + R_{\text{L}}) \tag{27-5}$$

Measurements. Measure primary and secondary winding resistances R_{p} and R_{s} using an ohmmeter, then use Equation 27-5 to compute the input impedance for the transformer. Compare to the value determined in (c) i.e., calculate the percent difference. If they differ, what are the likely sources of the difference?

$Z_{\text{in}} =$ _____

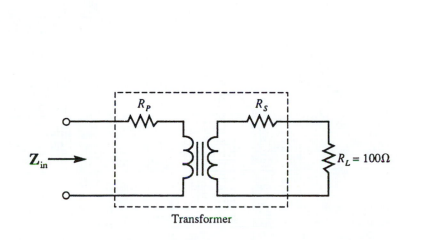

Figure 27-6 Transformer circuit with winding resistance included

PART D: Frequency Response of a Transformer

4. a. Using the audio transformer, re-connect the circuit of Figure 27-3, but add a 100 Ω load resistor between points c and d. Maintain the input voltage constant at 6 V (i.e., 12 V_{p-p}) and measure the output voltage and record in Table 27-2.

   b. Compute the ratio of output voltage to input voltage over the range of frequencies tested and plot as Figure 27-7.

   c. From the observed data, what conclusion can you draw about the validity of the ideal model?

f	V_{out}	V_{out}/V_{in}
100 Hz		
500 Hz		
1 kHz		
5 kHz		
10 kHz		
15 kHz		
20 kHz		

Table 27-2

PART E: Load Voltage Regulation

5. Ideally, a transformer will deliver constant output voltage. In reality, due to internal voltage drops, the output voltage falls as load current increases. In this test, we look at how badly the transformer voltage drops.

   a. Omitting the primary side voltmeter and observing suitable safety precautions, set up the circuit of Figure 27-2 and apply full rated voltage to the transformer.

   b. With a meter, measure the open circuit (no-load) voltage on the 12.6-V winding.

   $V_{NL} = $ _____

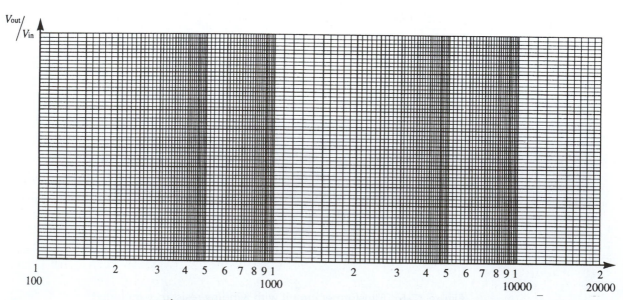

Figure 27-7   Frequency response for an audio transformer

c. Connect a load resistor that will draw full rated current (or nearly full-rated current). (If you use the Hammond 167L12 as recommended, you will need about 6 Ω, capable of dissipating 25 W. Since this is not a standard resistor, use a 10-Ω resistor instead. You will easily observe the regulation.) Measure output voltage under load. V_{FL} = _____

d. Compute the regulation of the transformer using the formula

$$\text{Regulation} = \left(\frac{V_{NL} - V_{FL}}{V_{FL}} \right) \times 100\%$$

PROBLEMS

6. An ideal transformer with a turns ratio of $a = 5$ has a load of $\mathbf{Z}_L = 10\,\Omega\,\angle30°$. Source voltage is $\mathbf{E} = 120\text{ V }\angle0°$.
   a. Compute the load voltage and load current as in Example 23-4 of the text.
   b. Using Equation 27-3, compute source current.
   c. Compute the input impedance $\mathbf{Z}_p$.
   d. Using $\mathbf{Z}_p$, compute source current and compare to the result obtained in (b).

7. A certain load requires a voltage that does not exceed 124 V or drop below 117 V. Two transformers are available, each with a no-load voltage of 120 V. Transformer 1 has a rated maximum regulation of ±3.% of its no-load voltage and Transformer 2 is rate 120-V ±3 V. Is either transformer suitable for the application? Perform an analysis to find out.

COMPUTER ANALYSIS

Ideal Transformers in Computer Analysis
MultiSIM: Use the virtual transformer (found in the *Basic* parts bin). Double click its symbol and set parameters as noted in Section 23.12 of the text to make it correspond to the ideal model. Note that Multi-SIM requires the use of

8. Using MultiSIM or PSpice and an ideal transformer with turns ratio $a = 5$ and $E = 120$ V, set up and emulate the circuit of Figure 27-3. (Omit the oscilloscope.) (PSpice users: Use source VSIN. Also place a large value resistor (say 100kΩ) across the secondary terminals of the transformer since PSpice does not like floating nodes.) Investigate waveforms, phase relationships and transformation ratios. Specifically, use the cursor to measure voltage magnitudes and in your analysis, use these to verify the turns ratio. Be sure to get printouts for both the in-phase and out-of-phase cases.

a ground on both sides of the transformer. **PSpice:** Use XFRM_LINEAR. (See boxed note in Section 23.12 of the text to refresh your memory on how to set it up to represent an ideal iron-core transformer.)

9. Consider a 10:1 iron-core transformer with $R_p = 16\ \Omega$, $L_p = 100$ mH, $R_s = 0.16\ \Omega$, $L_s = 1$ mH and a load consisting of a $10\ \mu$F capacitor in parallel with a $4\ \Omega$ resistor. The source is $120\ V\angle0°$.

a. Using MultiSIM or PSpice, solve for the load voltage. (This is full-load voltage.) Now solve for no-load voltage, then compute regulation.

b. Solve the above problem manually (with your calculator) and compare results.

FOR FURTHER INVESTIGATION AND DISCUSSION

Using the equivalent circuit of Figure 23-41 of the text and MultiSIM or PSpice, investigate the frequency response of an audio transformer. Use the following parameter values: $R_p = 1\ \Omega$, $L_p = 0.125$ mH, $R_s = 0.04\ \Omega$, $L_s = 0.005$ mH, $R_C = 500\ \Omega$, $L_m = 0.1$ H, turns ratio $a = 5$ and a speaker load of $8\ \Omega$, purely resistive. (Omit the stray capacitances). Use a 1 V AC source and run a frequency response curve from 10 Hz to 30 kHz. Submit a printout of your circuit plus the response curve. Write a short analysis of your investigation, describing in your own words the significance of what the curve shows in terms of the transformer's impact on sound quality in an audio system. Other points to cover include 1) over what range does the ideal model apply and why, and 2) what causes the curve to fall off at each end of the frequency spectrum?

Mutual Inductance and Loosely Coupled Circuits

LAB 28

OBJECTIVE

After completing this lab, you will be able to
- determine mutual inductance experimentally,
- measure mutual voltage and verify theoretically,
- compare calculated and measured results for coupled circuits.

EQUIPMENT REQUIRED

- ☐ Oscilloscope, dual channel
- ☐ Signal or function generator
- ☐ Inductance meter or bridge

COMPONENTS

- ☐ 455 kHz IF (*intermediate frequency*) transformer
- ☐ Resistor: 10-Ω, 1/4-W
- ☐ Capacitors: 1-μF, non-electrolytic (two required)

EQUIPMENT USED

Instrument	Manufacturer/Model No.	Serial No.
Oscilloscope		
Signal or function generator		
Inductance meter or bridge		

Table 28-1

TEXT REFERENCE

Section 23.9   LOOSELY COUPLED CIRCUITS
Section 23.10 MAGNETIC COUPLING IN NETWORK ANALYSIS
Section 23.11 COUPLED IMPEDANCE

DISCUSSION

If two coils with mutual coupling are connected in series so that their fluxes add as in Figure 28-1(a), their total inductance is

$$L_{T+} = L_1 + L_2 + 2M \tag{28-1}$$

where L_1 and L_2 are the self-inductances of coils *1* and *2* and M is the mutual coupling between them. If the coils are connected so that their fluxes subtract as in (b), total inductance is

$$L_{T-} = L_1 + L_2 - 2M \tag{28-2}$$

The mutual inductance between the coils can be found from the formula

$$M = (L_{T+} - L_{T-})/4 \tag{28-3}$$

(Since inductances L_1, L_2, L_{T+} and L_{T-} can be measured with a standard inductance meter, you can use Equation 28-3 to determine mutual inductance experimentally.) Self and mutual inductances are related by the formula

$$M = k\sqrt{L_1 L_2} \tag{28-4}$$

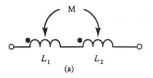

L_1 L_2
(a)

(a) Additive fluxes

where k is called the *coefficient of coupling*. For tightly coupled coils, $k = 1$; for loosely coupled coils, k is much less than 1.

Equations for Coupled Circuits. If coupled coils are connected as in Figure 28-2(a), their voltages and currents are related by the formulas

$$v_1 = R_1 i_1 + L_1 \frac{di_1}{dt} \pm M \frac{di_2}{dt} \tag{28-5}$$

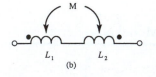

L_1 L_2
(b)

(b) Subtractive fluxes

$$v_2 = \pm M \frac{di_1}{dt} + R_2 i_2 + L_2 \frac{di_2}{dt} \tag{28-6}$$

where the sign of M is determined by the dot convention—i.e., if both currents enter (or leave) dotted terminals, then the sign to use with M is positive, while if one current enters a dotted terminal and the other leaves, the sign to use is negative. For sinusoidal excitation as shown in Figure 28-2(b), Equations 28-5 and 28-6 become

Figure 28-1 Coupled coils

$$\mathbf{V}_1 = (R_1 + j\omega L_1)\mathbf{I}_1 \pm j\omega M\mathbf{I}_2 \tag{28-7}$$

$$\mathbf{V}_2 = \pm j\omega M\mathbf{I}_1 + (R_2 + j\omega L_2)\mathbf{I}_2 \tag{28-8}$$

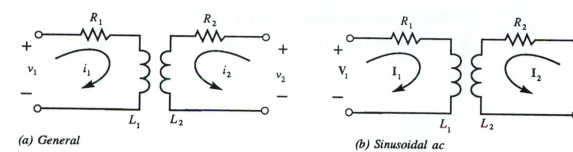

Figure 28-2   Coupled coils

MEASUREMENTS

For this lab, use a 455 kHz IF (*intermediate frequency*) transformer with coils wound on a ferrite core. (Note: Some IF transformers come with small capacitors connected. Make sure you disconnect them.)

PART A: Coil Parameters and Mutual Inductance

1. a. Measure the primary and secondary inductances and resist-ances using an *LRC* meter or impedance bridge.

L_1 = _____ R_1 = _____

L_2 = _____ R_2 = _____

b. Mark the coil ends for identification, then connect the primary and secondary coils in series as in Figure 28-1. Measure the induc-tance of the series connection, then reverse the connections and measure again.

L_T+ _____ L_T- = _____

Use Equation 28-3 to compute *M*. M = _____

Use Equation 28-4 to compute *k*. k = _____

Is this transformer loosely coupled or tightly coupled? _____

c. Use the results of Test 1(b) to determine the dotted ends for the coils. (Mark the dotted ends as you will need this information later.) Describe how you arrived at this conclusion.

_____

_____

PART B: Mutually Induced Voltage

2. Measure the value of the 10-Ω sensing resistor R_S, then assemble the circuit as in Figure 28-3.

   $R_S =$ _____

   a. Set the source voltage to 4 V_p (8 V_{p-p}) at f = 1 kHz. Measure V_S, magnitude and angle (using the oscilloscope) and record below. Compute input current I_1, magnitude, and angle. (You can leave the results in peak volts and amps, rather than convert to rms.)

   $V_S =$ _____ ∠ $I_1 =$ _____ ∠

   b. With Probe 1 still measuring e, move Probe 2 to measure secondary voltage v_2. Measure magnitude and angle.

   $V_2 =$ _____ ∠

   c. Using the measured values of E, R_1, R_S and L_1, compute current I_1. Thus,

   $$I_1 = \frac{E}{R_1 + R_S + j\omega L_1}$$

   Compare to the result measured in (a).

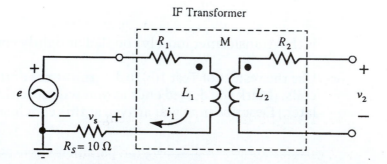

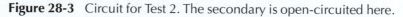

Figure 28-3   Circuit for Test 2. The secondary is open-circuited here.

d. Compute V_2 from the relationship

$V_2 = j\omega M I_1 =$

Compare to the result measured in (b).

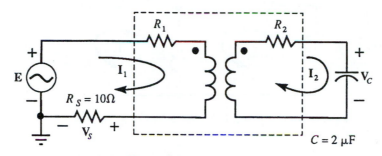

Figure 28-4   Circuit for Test 3

PART C: Coupled Circuit Equations

3. Add 2 µF of capacitance to the secondary circuit as in Figure 28-4.
   a. Set e to 4 V_p (8 V_{p-p}). Carefully measure V_S and V_2 as before.

   $V_S =$ _____ $\angle$ $V_2 =$ _____ $\angle$

   Compute current $I_1 = V_S/R_1 =$ _____ $\angle$

   b. Write mesh equations for this circuit and solve for I_1 and I_2. Using the computed value of I_2, calculate V_2. Compare I_1 and V_2 to the values measured in 3(a). How well do they agree?
   c. Solve for I_1 using the coupled impedance equation

   $$Z_{in} = Z_1 + \frac{(\omega M)^2}{Z_2 + Z_L}$$

   Compare to the measured value.

COMPUTER ANALYSIS

Note re: MultiSIM
At the time of writing, MultiSIM has no simple way to handle loosely coupled circuits. Therefore, omit #4 and #5.

Use XFRM_LINEAR for the following.

4. Consider Figure 28-3. Using the parameter values that you measured in Part A, simulate this circuit using PSpice and determine V_2. Compare to the measured value.

5. Simulate the circuit of Figure 28-4 using PSpice. Using the parameter values that you measured in Part A, determine I_2 and V_2. Compare to the measured values.

PROBLEM

6. Write equations for the circuit of Figure 28-5 and solve for I_1, I_2 and V_2.

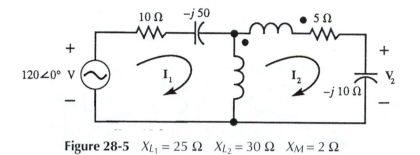

Figure 28-5 $X_{L_1} = 25\ \Omega$ $X_{L_2} = 30\ \Omega$ $X_M = 2\ \Omega$

FOR FURTHER INVESTIGATION AND DISCUSSION

Use PSpice to simulate the circuit of Figure 28-5 at $f = 60$ Hz. Your written report should explain how you used XFRM_LINEAR to model the circuit coupling shown, as well as anything else you had to do to solve this problem.

The Basic Power Supply: Rectification, Filtering and Regulation

OBJECTIVES

After completing this lab, you will be able to
- Construct half and full wave rectifier circuits,
- Measure and analyze filtered waveforms,
- Measure load regulation,
- Use a 3-terminal regulator to regulate dc supplies,
- Calculate the ripple rejection of a regulated supply.

EQUIPMENT REQUIRED

☐ DMM
☐ Two-channel oscilloscope

COMPONENTS

☐ Class 2 transformer 12 VAC, 500 mA
☐ 1N4004 diodes or equivalent (4 required)
☐ Resistors (1 each): 560-Ω, 1/2-W; 100-Ω, 1-W; 220-Ω, 1-W; 51-Ω, 10-W
☐ 470 μF electolytic capacitor (1 only)
☐ 7805 voltage regulator and data sheet

CAUTION

Be sure to turn off the power before you assemble or make changes to your circuit.

TEXT REFERENCE

Circuit Analysis with Devices: Theory and Practice
Section 27.7   HALF AND FULL WAVE RECTIFIER CIRCUITS
Section 27.8   POWER SUPPLY FILTERING

DISCUSSION

Diodes are designed for various purposes—signal diodes for high frequency, low power applications, rectifier diodes for higher current power supply applications, and so on. Because we are dealing with power supplies here, we require rectifier diodes and have chosen the popular 1N4004.

Rectification produces pulsating dc. Such crude dc, however, is unsuitable for most applications and requires smoothing. The simplest approach to smoothing is to place a large value capacitor across the load. This capacitor (known as a filter) helps level out the waveform, but some ripple remains.

In addition to the ripple problem, the output voltage of a power supply falls off as load current is increased. This change from no load voltage (V_{NL}) to full load voltage (V_{FL}) is termed regulation and is defined as

$$\text{Regulation} = \frac{V_{NL} - V_{FL}}{V_{FL}} \times 100\% \qquad (29\text{-}1)$$

In practice, you want a power supply with small ripple and small regulation. A simple way to improve performance is to add an inexpensive 3-terminal regulator to your circuit. This reduces both ripple and regulation. The ability of a regulator to reduce ripple (referred to as *ripple rejection*), is defined as the degree to which the regulator is able to prevent ripple voltage from passing from its input to its output. It is expressed in dB as

$$[\text{ripple rejection}]_{dB} = 20 \log\left(\frac{V_{\text{ripple(in)}}}{V_{\text{ripple(out)}}}\right) \qquad (29\text{-}2)$$

In this lab, we use the 7805, a popular, inexpensive and widely used solid-state, 3-terminal voltage regulator.

MEASUREMENTS

PART A: Half-Wave Rectifier Circuit and Filtering

1. Construct the half-wave circuit of Figure 29-1(a).
2. Connect an oscilloscope across the load resistor. Set the scope to its dc mode and sketch load voltage v_L in Figure 29-2(a). Be sure to show key voltage and time measurement data. Carefully measure peak voltage and record.

$V_m = $ _____

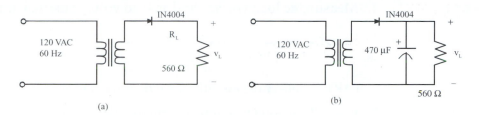

Figure 29-1 Half-wave rectification

3. Add a filter capacitor (see Caution) as in Figure 29-1(b). Connect both oscilloscope probes across the resistor to measure load voltage. Set Channel 1 to dc coupling and Channel 2 to ac coupling.
   a. Sketch the dc waveform in Figure 29-2(b), carefully labeling voltage and time axes.

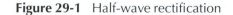

(a) **Unfiltered**   *(b)* **Filtered**

Figure 29-2

   b. Expand the vertical scale on Channel 2 and measure peak-to-peak ripple and ripple frequency and sketch the waveform in Figure 29-3 and carefully label.

$V_{r_{pp}} =$ _____

$f_{ripple} =$ _____

Figure 29-3 AC ripple

4. Measure dc load voltage and ac load voltage using your DMM.

    $V_{L(dc)} =$ _____ $V_{L(ac)} =$ _____

PART B: Full-Wave Rectifier Circuit and Filtering

5. Using $R_L = 560 \ \Omega$, assemble the full-wave bridge circuit of Figure 29-4, temporarily omitting the capacitor.
    a. Using the oscilloscope (dc coupled), observe and sketch (in Figure 29-5) the load voltage.
    b. Add the capacitor (see Caution) and using dc coupling, sketch the filtered waveform on the same graph. Label voltage magnitudes.
    c. Change to ac coupling and measure the ripple frequency,

        $f_{ripple} =$ _____

6. Make load measurements as follows and record in Table 29-1:
    a. With an ac-coupled oscilloscope, measure the peak-to-peak ripple voltage and record.

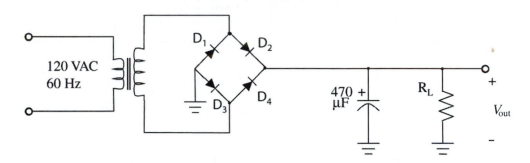

Figure 29-4 Full-wave bridge

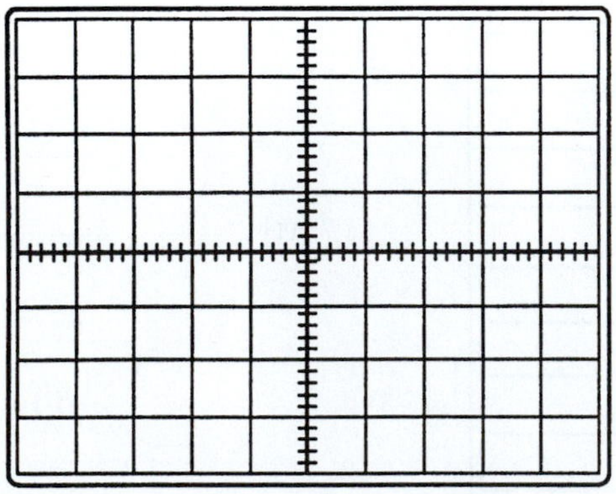

Figure 29-5

b. With a DMM set to dc, measure load voltage $V_{out(dc)}$ and record. Compute load current I_{dc} and record.
7. Repeat Step 6 using a 220-Ω resistor.
8. Repeat Step 6 using the 100-Ω load. **Caution: The resistor may become quite hot.**
9. Repeat Step 6 using the 51-Ω load. **Caution: The resistor will become quite hot.**

R_L	Measured		I_{dc}	Computed	
(ohms)	$V_{r_{p-p}}$	$V_{out(dc)}$	(mA)	$V_{r_{p-p}}$	$V_{out(dc)}$
560					
220					
100					
51					

Table 29-1

PART C: 3-Terminal Voltage Regulator Circuit

10. Using the 560 Ω-resistor and a 7805, 5-V regulator, set up the circuit of Figure 29-6.
11. Make measurements as follows and record results in Table 29-2.
    a. Set your DMM to dc and measure load voltage $V_{out(dc)}$. Calculate load current I_{dc}.
    b. Set your oscilloscope to ac coupled and measure the output ripple voltage $V_{r_{p-p}}$.
12. Repeat Step 11 using the 220-Ω resistor.
13. Repeat Step 11 using the 100-Ω resistor.
14. Repeat Step 11 using the 51-Ω resistor.

R_L	Measured		I_{dc}
(ohms)	$V_{r_{p-p}}$	$V_{out(dc)}$	(mA)
560			
220			
100			
51			

Table 29-2

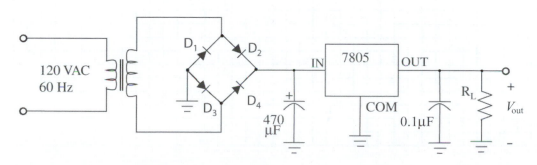

Figure 29-6   Regulated supply

15. With the 51-Ω resistor still connected, use the oscilloscope to measure the peak-to-peak ripple voltage at the input to the regulator.

$V_{\text{ripple(input)}}$ = _____

CALCULATIONS, DISCUSSION AND CONCLUSIONS

16. Using the techniques of Chapter 27, Section 27.8 of the text and the value of V_m measured in Step 2, compute peak-to-peak ripple voltage and dc load voltage for the half-wave filtered waveform and compare to your measured results.

17. In Step 3 you measured the ac component of the load voltage (the ripple) using an oscilloscope and in Step 4, you measured it with a DMM set to AC. Based on your results, comment on how useful a DMM is in determining ripple voltage and why.

18. Using the techniques of Chapter 27, Section 27.8 of the text and the value of V_m measured in Step 5, compute peak-to-peak ripple voltage and dc load voltage for the full-wave filtered waveform for each value of load resistance and record in Table 29-1. How do these compare to your measured results?

19. Using the graph of Figure 29-7 and your data from Steps 6 to 9, plot load dc voltage versus load current for the unregulated supply.

20. From the graph of Figure 29-7, compute regulation.

21. Repeat 19 and 20 for the regulated supply, using the graph of Figure 29-8. Compare the percent regulation for the unregulated supply against that for the regulated supply.

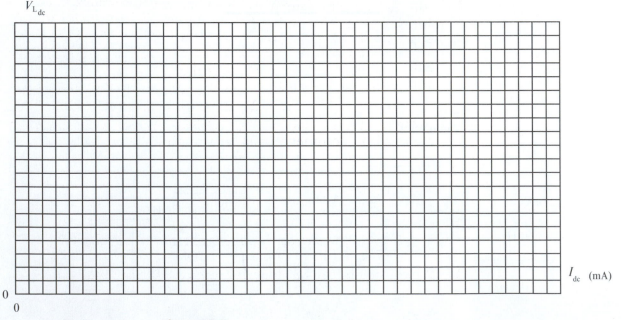

Figure 29-7

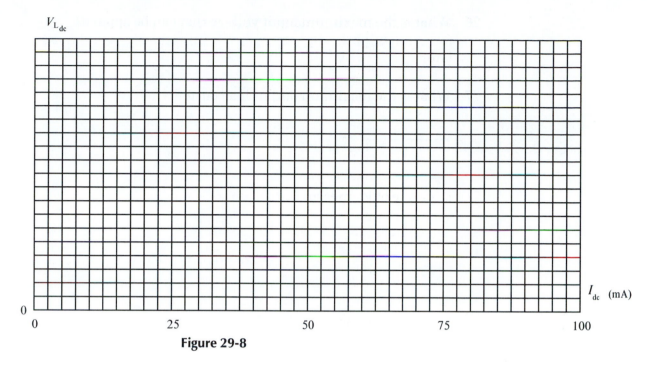

Figure 29-8

22. For the regulated supply, compute ripple rejection at a load current of $I_{dc} \approx 100$ mA based on your measured data.

COMPUTER ANALYSIS

23. Using either MultiSIM or PSpice, determine V_{dc} and peak-to-peak ripple for the circuit of Figure 29-4 with $R_L = 51\ \Omega$. Hint: For simplicity, omit the transformer and drive the rectifier combination directly from an AC source whose voltage E_m is two diode drops greater than V_m as measured in Step 5a. How do these results compare to the measured results?

24. To investigate how a filter capacitor affects ripple, run simulations with $C = 100\ \mu F$, $C = 220\ \mu F$, $C = 330\ \mu F$ and $C = 1000\ \mu F$ for the case $R_L = 51\ \Omega$, and plot ripple voltage versus C. Submit your graph with a commentary on your findings.

PROBLEMS

Refer to the manufacturer's data sheet to answer the following.

25. What is the minimum required input voltage for the 7805?

_____ V

26. What is the maximum input voltage that can be applied?

_____ V

27. What is the maximum output load current?

_____ A

NAME _____

DATE _____

CLASS _____

Introduction to Transistor Amplifiers

Objectives

After completing this lab, you will be able to
- calculate the operating point of a transistor,
- use a digital multimeter to measure the operating point of a transistor,
- calculate the theoretical voltage gain, input impedance, and output impedance of a CE amplifier,
- use an oscilloscope to measure the voltage gain, input impedance, and output impedance of a CE amplifier.

Equipment Required

- ☐ Digital multimeter (DMM)
- ☐ dc power supply
- ☐ Signal generator (sinusoidal function generator)
- ☐ Two-channel oscilloscope
- ☐ *Note:* Record this equipment in Table 30-1.

Components

- ☐ Transistors: 2N3904 npn
- ☐ Resistors: 180-Ω, 820-Ω, 1-kΩ(two), 1.5-kΩ, 2.2-kΩ, 10-kΩ, 3.3-kΩ.
- ☐ Capacitors: 10-µF electrolytic (two), 470-µF electrolytic.

EQUIPMENT USED

Instrument	Manufacturer/Model No.	Serial No.
DMM		
dc power supply		
Oscilloscope		
Signal Generator		

Table 30-1

TEXT REFERENCE

Circuit Analysis with Devices: Theory and Practice
Section 28.6   TRANSISTOR BIASING
Section 29.2   BJT SMALL-SIGNAL MODELS
Section 29.4   THE COMMON-EMITTER AMPLIFIER

DISCUSSION

The universal-bias transistor circuit shown in Figure 30-1 is the most commonly used transistor amplifier circuit. When correctly

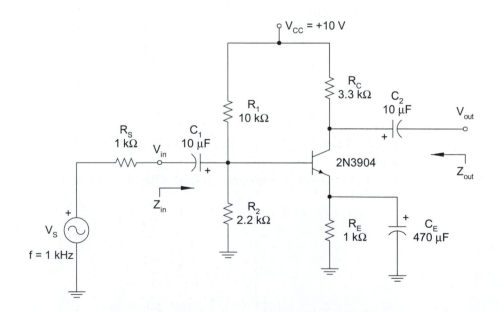

Figure 30-1   Universal bias CE amplifier

designed, the universal-bias amplifier is very stable. This means that the amplifier has highly predictable results even when the transistor beta changes due to variation in the manufacturing process or due to temperature.

In this lab you will measure the operating point (also called the quiescent or Q-point) of a transistor. Once the transistor is biased correctly, you will apply a sinusoidal voltage at the input of the transistor amplifier and observe that the output signal will be amplified by the transistor circuit.

CALCULATIONS

1. Refer to the circuit of Figure 30-1. Calculate I_C and V_{CE} at the operating point. Show your work in the space provided below.

I_{CQ}	mA

V_{CEQ}	V

MEASUREMENTS

2. Construct the circuit shown in Figure 30-1. Ensure that you place the electrolytic capacitors correctly.
3. Measure and record the following dc voltages using a dc multimeter (DMM).

V_B	V

V_E	V

V_C	V

4. Use the measurements of Step 3 to calculate the operating point of the transistor. Enter your values in the space provided.

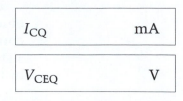

I_{CQ}	mA

V_{CEQ}	V

If your measured results are not within 10% of the calculated values of Step 1, call your instructor to determine the source of the problem.

5. With the oscilloscope connected to v_{in}, adjust the signal generator to apply a sinusoidal signal of 20 m$V_{p\text{-}p}$ and 1 kHz to v_{in}.
6. Measure and record the output voltage v_{out}.

v_{out}	$V_{p\text{-}p}$

7. Calculate the voltage gain of the amplifier and enter the result in the space provided.

A_v

8. Without adjusting the signal generator, measure and record the voltage at the output of the signal generator. Use your measurement to calculate the input impedance of the amplifier. Show your calculations in the space provided.

v_{gen}	$V_{p\text{-}p}$

z_{in}	Ω

9. Without adjusting the signal generator, connect a load resistor of 4.7 kΩ between v_{out} and ground in the circuit of Figure 30-1. Measure the new value of output voltage. (You should find that the voltage has decreased from the value found in Step 6.) Use your measurement to calculate the output impedance of the circuit.

V_{out}	V_{p-p}

z_{out}	Ω

10. Modify the circuit as illustrated in Figure 30-2.

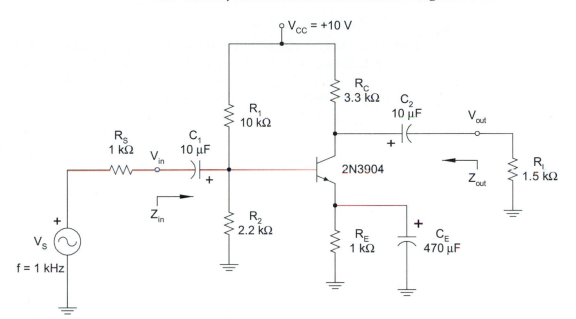

Figure 30-2   Universal bias CE amplifier with load resistor

11. Measure and record the following dc voltages using a dc multimeter (DMM).

V_B	V

V_E	V

V_C	V

If your measured values have changed from those found in Step 3, call your instructor to determine the source of the problem.

12. With the oscilloscope connected to v_{in}, adjust the signal generator to apply a sinusoidal signal of 20 mV_{p-p} and 1 kHz to v_{in}.

13. Measure and record the output voltage v_{out}.

v_{out}	V_{p-p}

14. Calculate the voltage gain of the amplifier and enter the result in the space provided.

A_v

15. Without adjusting the signal generator, measure and record the voltage at the output of the signal generator. Use your measurement to calculate the input impedance of the amplifier. Show your calculations in the space provided.

v_{gen}	V_{p-p}

z_{in}	Ω

16. Did the value of input impedance change from the value found in Step 8. Circle the correct answer.

yes/no

17. Modify the circuit as illustrated in Figure 30-3.

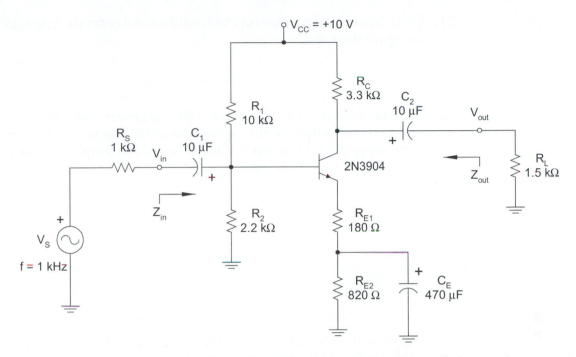

Figure 30-3   Universal bias CE amplifier with partially bypassed emitter resistor

18. Measure and record the following dc voltages using a dc multimeter (DMM).

V_B	V

V_E	V

V_C	V

If your measured values have changed from those found in Step 3, call your instructor to determine the source of the problem.

19. With the oscilloscope connected to v_{in}, adjust the signal generator to apply a sinusoidal signal of 20 m$V_{p\text{-}p}$ and 1 kHz to v_{in}.
20. Measure and record the output voltage v_{out}.

v_{out}	$V_{p\text{-}p}$

21. Calculate the voltage gain of the amplifier and enter the result in the space provided.

22. Without adjusting the signal generator, measure and record the voltage at the output of the signal generator. Use your measurement to calculate the input impedance of the amplifier. Show your calculations in the space provided.

V_{gen}	$V_{p\text{-}p}$

z_{in}	Ω

23. Did the value of input impedance change from the value found in Step 8. Circle the correct answer.

yes/no

Problem:

Refer to the circuit of Figure 30-3. Assume that $\beta_{ac} = 150$.
  a. Sketch the ac equivalent circuit in the space provided below.

b. Calculate the voltage gain A_v, input impedance z_{in}, and output impedance z_{out}. Show your work in the space provided.

z_{in}	Ω

z_{out}	Ω

The Differential Amplifier

Objectives

After completing this lab, you will be able to
- measure single-ended voltage gain of a differential amplifier,
- recognize the phase difference between the input and each of the outputs of a differential amplifier,
- measure double-ended voltage gain of a differential amplifier,
- measure the common-mode gain of a differential amplifier,
- calculate the common-mode rejection ratio (CMRR) of a differential amplifier as both a ratio and in dB.

Equipment Required

☐ Digital multimeter (DMM)
☐ Dual power supply
☐ Signal generator (sinusoidal function generator)
☐ Two-channel oscilloscope
Note: Record this equipment in Table 31-1.

Components

☐ Transistors: 2N3904 npn (three)
☐ Resistors: 1-kΩ (three), 2.2-kΩ (two), 2.7-kΩ (two),
☐ Diodes: 1N4735A zener diode (V_Z = 6.2 V)
☐ Capacitors: 10-µF electrolytic (two).

EQUIPMENT USED

Instrument	Manufacturer/Model No.	Serial No.
DMM		
Dual power supply		
Oscilloscope		
Signal Generator		

Table 31-1

TEXT REFERENCE

Section 30.2 THE DIFFERENTIAL AMPLIFIER AND COMMON-MODE SIGNALS

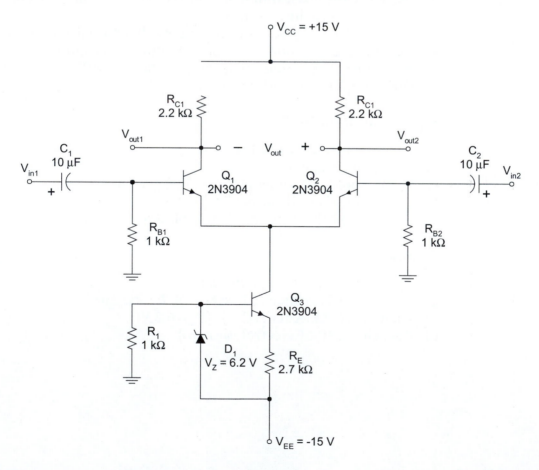

Figure 31-1 Differential amplifier

MEASUREMENTS

1. Build the circuit shown in Figure 31-1.
2. Measure and record the following voltages with a DMM.

V_{B1} _____ V	V_{B2} _____ V	V_{B3} _____ V
V_{C1} _____ V	V_{C2} _____ V	V_{C3} _____ V
$V_{E1} = V_{E2} = V_{C3}$ _____ V		

Using the measurements of Step 2, calculate the following:

I_{C1} _____ mA	I_{C2} _____ mA	I_{C3} _____ mA
V_{CE1} _____ V	V_{CE2} _____ V	V_{CE3} _____ V

Single-ended differential amplifier

4. Connect v_{in2} to ground. Apply a suitable 1-kHz sinusoidal signal to v_{in1}. Measure v_{in1} on CH1 of the oscilloscope and v_{out1} and v_{out2} on CH2. Record all values.

V_{in1} _____ V$_{p-p}$	V_{out2} _____ V$_{p-p}$	V_{out2} _____ V$_{p-p}$

5. Are v_{in1} and v_{out1} in phase or out of phase? _____

6. Are v_{in1} and v_{out2} in phase or out of phase? _____

7. Calculate the theoretical single-ended voltage gain of the amplifier.

$$A_{d(S.E.)} = \frac{v_{out}}{v_{in1}} = -\frac{R_C}{2r_e} \qquad (31\text{-}1)$$

$A_{d(S.E.)}$

8. Calculate the actual single-ended voltage gain of the amplifier from the measurements of Step 4.

$A_{d(S.E.)}$

Double-ended differential amplifier

9. Without readjusting the signal generator voltage or frequency, connect CH1 of the oscilloscope to v_{out1} and CH2 to v_{out2}. Place the oscilloscope into its CH2 – CH1 mode and measure the differential voltage at the output.

v_{out2}	$V_{p\text{-}p}$

10. Calculate the theoretical differential voltage gain of the amplifier.

$$A_{d(D.E.)} = \frac{v_{out}}{v_{in1}} = -\frac{R_C}{r_e} \qquad (31\text{-}2)$$

$A_{d(D.E.)}$

11. Calculate the actual differential voltage gain of the amplifier from the measurement of Step 9.

$A_{d(D.E.)}$

12. Leave the two channels of the oscilloscope connected as in Step 11. With one end of the signal generator connected to the signal ground, connect the other terminal to both inputs, v_{in1} and v_{in2}. Increase the output of the generator until you are able to measure a suitable common-mode output voltage, v_{out}. Record your value.

v_{out2}	$V_{p\text{-}p}$

13. Return your oscilloscope to normal operation and measure the common-mode input voltage. Record your value.

v_{in2}	$V_{p\text{-}p}$

14. Use the results of Steps 12 and 13 to calculate the common-mode voltage gain.

A_{cm}

15. Calculate the common-mode rejection ratio.

$$CMRR = \frac{A_{d(D.E.)}}{A_{cm}} \qquad (31\text{-}3)$$

CMRR

16. Calculate the common-mode rejection ratio in decibels.

$$[\text{CMMR}]_{\text{dB}} = 20 \log \text{CMRR} \qquad (31\text{-}4)$$

CMRR	dB

COMPUTER SIMULATION

17. Use either MultiSIM or PSpice to simulate the circuit of Figure 31-1. Apply a 10 m$V_{\text{p-p}}$ signal to v_{in1} and connect v_{in2} to ground. Observe and record the voltages observed at v_{out1} and v_{out2}.

V_{in1} 10 m$V_{\text{p-p}}$	v_{out2} $V_{\text{p-p}}$	v_{out2} $V_{\text{p-p}}$

18. Use the measurements of Step 17 to calculate both the single-ended voltage gain and the double-ended voltage gain of the amplifier. Record your calculations in the space provided.

$A_{\text{d(S.E.)}}$

$A_{\text{d(D.E.)}}$

19. Compare the results observed in the computer simulation to the actual measured values. Offer an explanation for any discrepancy.

Introduction to Op-Amps

Objectives

After completing this lab, you will be able to
* calculate voltage gain, input impedance, and output impedance of an inverting operational amplifier (op-amp) circuit at mid frequency,
* measure and plot the voltage gain as a function of frequency for an inverting op-amp circuit,
* measure the upper cutoff frequency of an op-amp and compare the result to that obtained from the manufacturer's specification,
* sketch the frequency response of an inverting op-amp circuit on semi-log graph paper,
* observe and measure the operation of an op-amp adder circuit.

Equipment Required

☐ Digital multimeter (DMM)
☐ Dual Power Supply
☐ Signal generator (sinusoidal function generator)
☐ Two-channel oscilloscope
   Note: Record this equipment in Table 32-1.

Components

☐ Op-Amps: 741C op-amp (8-pin package)
☐ Resistors: 150-Ω (two) , 1.0-kΩ, 1.2-kΩ, 10-kΩ, 75-kΩ (three).

EQUIPMENT USED

Instrument	Manufacturer/Model No.	Serial No.
DMM		
Dual power supply		
Oscilloscope		
Signal Generator		

Table 32-1

TEXT REFERENCE

Circuit Analysis with Devices: Theory and Practice
Section 30.1   INTRODUCTION TO THE OPERATIONAL
AMPLIFIER
Section 30.3   NEGATIVE FEEDBACK
Section 30.4   THE INVERTING AMPLIFIER

CALCULATIONS

1. Given the circuit shown in Figure 32-1, calculate A_v, z_{in}, and z_{out}.

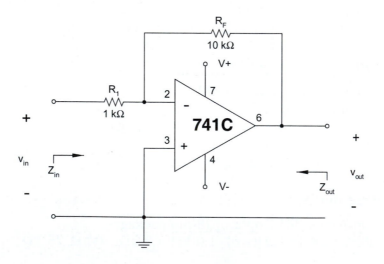

Figure 32-1   Inverting amplifier

A_v		z_{in}	Ω	z_{out}	Ω

MEASUREMENTS

2. Adjust the dual power supply to provide ±15 V with respect to the common (ground) point. Connect the power to the op-amp as illustrated in Figure 32-1. The pins shown in the illustration are for the 8-pin DIP package. If you use a different package, you will need to obtain the pin numbers from the manufacturer's specification.

3. Adjust the sinusoidal generator to provide an input voltage, $v_{in} = 0.1$ V$_{p-p}$ at a frequency of 1 kHz. Measure the output voltage, V_{out}.

| V_{out1} | V$_{p-p}$ |

4. Use the measured values of Step 3 to calculate the mid-frequency voltage gain of the op-amp circuit.

| A_v |

5. Measure the input impedance and the output impedance of the amplifier.

| z_{in} | Z_{out} | Ω |

6. Keeping the amplitude of the signal constant, adjust the frequency of the signal generator and measure the output voltage at each of the indicated frequencies. Use your measurements to calculate voltage gain and gain in decibels at each frequency. Calculate the voltage gain in decibels using the following expression:

$$\left[A_v\right]_{dB} = 20 \log \left| \frac{v_{out}}{v_{in}} \right| \tag{32-1}$$

Enter your data in Table 32-2 and show one set of calculations in the space provided.

7. Adjust the frequency of the generator so that the output voltage is exactly 0.707 of the value measured in Step 2. Measure and record the upper cutoff frequency of the amplifier.

| f_H | Hz |

Frequency	v_{out} (V$_{p-p}$)	A_v	$[A_v]_{dB}$
10 Hz			
20 Hz			
50 Hz			
100 Hz			
200 Hz			
1 kHz			
2 kHz			
5 kHz			
10 kHz			
20 kHz			
50 kHz			
100 kHz			
200 kHz			
500 kHz			
1 MHz			

Table 32-2   Frequency response measurements

Sample calculations:

8. Plot the data of Table 32-2 as a semi-logarithmic graph.
9. Refer to the manufacturer's specifications and determine the gain-bandwidth product for the 741C op-amp.

Gain-bandwidth product   Hz

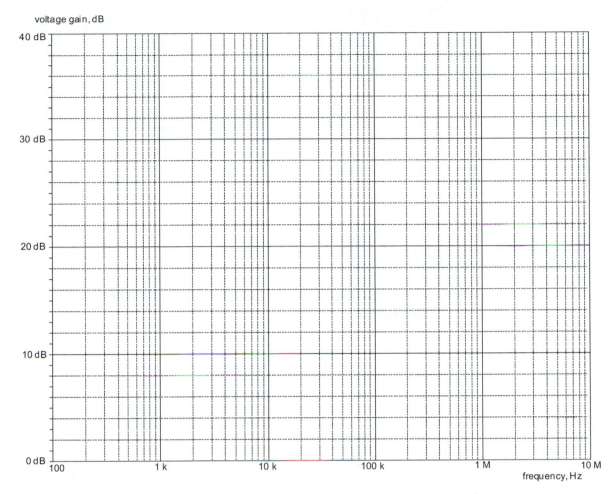

Graph 32-1   Frequency response of an inverting amplifier

10. Use the gain-bandwidth product and the mid-frequency gain of the amplifier to determine the theoretical upper cutoff frequency of the amplifier.

f_H _____ Hz

How well does the actual bandwidth compare to the theoretical bandwidth?

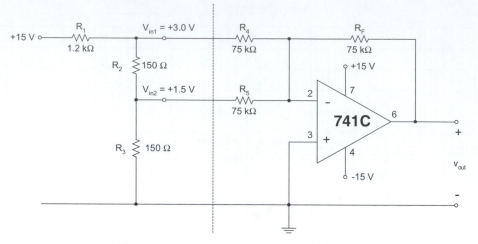

Figure 32-2   Inverting two-input adder

11. Given the circuit shown in Figure 32-2, predict the output voltage, V_{out}.

V_{out}	V

12. Use a digital multimeter to measure the output voltage of the circuit of Figure 32-2.

V_{out}	V

How well does the actual output voltage compare to the theoretical output?

FOR FURTHER INVESTIGATION AND DISCUSSION

How does the bandwidth (upper cutoff frequency) of an op-amp circuit amplifier depend on the gain of the amplifier? To aid in your discussion, do the following:

a. Use MultiSIM or PSpice to simulate the circuit shown in Figure 32-1. Use the Bode plotter (MultiSIM) or the Probe processor (PSpice) to determine the frequency response of the amplifier. Use cursors to determine the upper cutoff frequency of the amplifier.

b. Modify the circuit of Figure 32-1 by letting $R_F = 20$ kΩ, and repeat Part a.

<table>
<tr><td>

LAB

33

</td></tr>
</table>

Non-Inverting Op-Amps

Objectives

After completing this lab, you will be able to
- calculate voltage gain of a non-inverting operational amplifier (op-amp) circuit at mid frequency,
- calculate input impedance and output impedance of a non-inverting op-amp circuit,
- measure voltage gain of a non-inverting op-amp circuit,
- measure the upper cutoff frequency of a non-inverting amplifier,
- measure loading effect with and without a buffer circuit.

Equipment Required

☐ Digital multimeter (DMM)
☐ Dual power supply
☐ Signal generator (sinusoidal function generator)
☐ Two-channel oscilloscope
   Note: Record this equipment in Table 33-1.

Components

☐ Op-Amps: 741C op-amp (8-pin package)
☐ Resistors: 1.0-kΩ (two), 9.1-kΩ, 10-kΩ, 10-MΩ (two),
☐ 10-kΩ potentiometer.

EQUIPMENT USED

Instrument	Manufacturer/Model No.	Serial No.
DMM		
Dual power supply		
Oscilloscope		
Signal Generator		

Table 33-1

TEXT REFERENCE

Circuit Analysis with Devices: Theory and Practice
Section 30.5   THE NON-INVERTING AMPLIFIER

CALCULATIONS

1. Given the circuit shown in Figure 33-1, calculate A_v, z_{in}, and z_{out}.

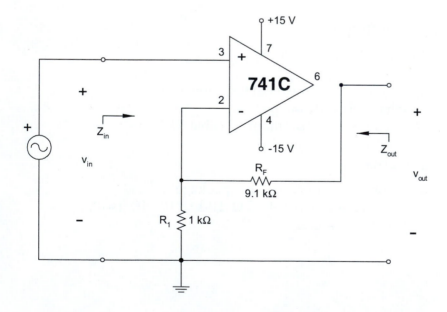

Figure 33-1   Non-inverting amplifier

A_v		z_{in}	Ω	z_{out}	Ω

MEASUREMENTS

2. Adjust the dual power supply to provide ±15 V with respect to the common (ground) point. Connect the power to the op-amp as illustrated in Figure 33-1. The pins shown in the illustration are for the 8-pin DIP package. If you use a different package, you will need to obtain the pin numbers from the manufacturer's specification.

3. Adjust the sinusoidal generator to provide an input voltage, $v_{in} = 0.1\ V_{p-p}$ at a frequency of 1 kHz. Measure the output voltage, V_{out}.

V_{out1}	V_{p-p}

4. Use the measured values of Step 3 to calculate the mid-frequency voltage gain of the op-amp circuit.

A_v

5. Modify the circuit by adding only the 10-MΩ resistor at the input as shown in Figure 33-2. Be careful that you do not readjust the generator voltage. Since an oscilloscope connected at the input of the op-amp will load the circuit you will need to take measurements at the output. Measure and record the output voltage with the 10-MΩ resistor in the circuit.

V_{out2}	V_{p-p}

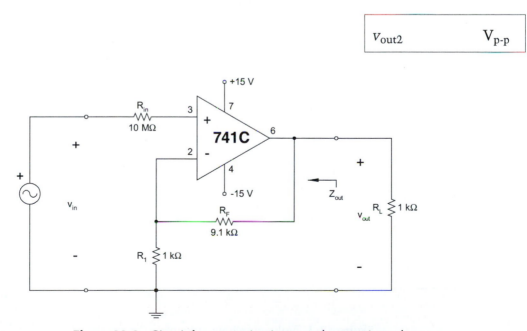

Figure 33-2   Circuit for measuring input and output impedance

6. Use Equation 33-1 to calculate z_in using the voltage measurements of Steps 3 and 5.

$$z_\text{in} = \left(\frac{V_\text{out(no load)}}{V_\text{(outloaded)} - V_\text{out(no load)}} \right) R_\text{in}$$ (33-1)

z_in	Ω

7. Modify the circuit by adding a 1-kΩ resistor at the output as shown in Figure 33-2. Measure the resulting output voltage and use this value as well as the value from Step 3 to calculate the output impedance of the amplifier.

v_out2	$V_\text{p-p}$

8. Construct the voltage follower circuit shown in Figure 33-3 and adjust the sinusoidal generator to provide an input voltage, $v_\text{in} = 0.1\ V_\text{p-p}$ at a frequency of 1 kHz. Measure the output voltage, v_out.

v_out	$V_\text{p-p}$

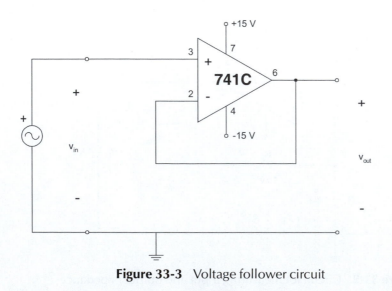

Figure 33-3 Voltage follower circuit

9. Calculate the mid-frequency voltage gain of the op-amp circuit of Figure 33-3.

10. Connect a potentiometer to the +15V as shown in the left-hand side of Figure 33-4 (non-loaded) and adjust the center terminal to provide 10 V as measured with a DMM. **Do not readjust the value of the potentiometer again.**

11. Now place a load resistor, R_L = 10 kΩ across the terminals of the potentiometer. Measure and record the resulting output voltage, V_{out}. You should have observed that the voltage went down. Explain why the measurements are different.

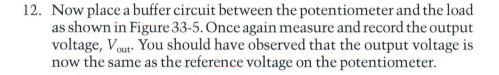

12. Now place a buffer circuit between the potentiometer and the load as shown in Figure 33-5. Once again measure and record the output voltage, V_{out}. You should have observed that the output voltage is now the same as the reference voltage on the potentiometer.

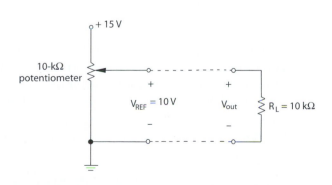

Figure 33-4 Loading effect of a potentiometer

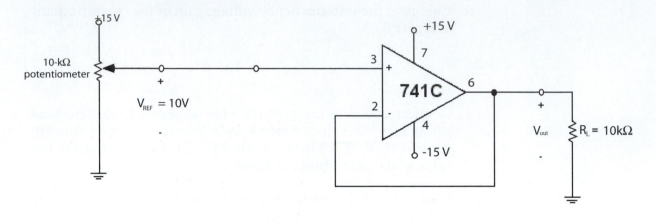

Figure 33-5   Application of a voltage follower

FOR FURTHER INVESTIGATION AND DISCUSSION

13. Use MultiSIM or PSpice to simulate the circuit of Figure 33-1. Determine the voltage gain of this circuit and compare the predicted value to your actual measured value of Step 4.

A_v [   ]

14. Use MultiSIM or PSpice to simulate the circuit of Figure 33-3. Determine the voltage gain of this circuit and compare the predicted value to your actual measured value of Step 9.

A_v [   ]

15. A standard digital multimeter has an input resistance of 10 MΩ when used to measure voltage. For normal applications, the high input resistance of the meter will not load a circuit. This however, is not true the circuit resistance is very large. Explain how a buffer circuit can be used to measure voltage in a circuit having very high resistances. Use a simple schematic to help in your discussion.

Op-Amp Applications

Objectives

After completing this lab, you will be able to
- calculate and measure the voltage and voltage gain for a balanced line driver,
- properly use an op-amp to build and connect a differential amplifier (receiver circuit) to a balanced line,
- correctly connect a common-mode signal to a differential amplifier and measure the common-mode voltage gain,
- measure the differential voltage gain of a differential amplifier,
- calculate the common-mode rejection ratio (CMRR) of an amplifier,
- predict and measure the frequency response of an integrator circuit,
- plot the frequency response of an integrator on semi-logarthmic graph paper,
- measure the output of an integrator circuit when a square wave is applied to the input circuit.

Equipment Required

☐ Digital multimeter (DMM)
☐ Dual power supply
☐ Signal generator (sinusoidal function generator)
☐ Two-channel oscilloscope
   Note: Record this equipment in Table 34-1.

Components

☐ Op-Amps: 741C op-amp (two), TL071 op-amp
☐ Resistors: 5.1-kΩ (three), 9.1-kΩ, 10-Ω (four), 47-kΩ, 51-kΩ, 100-kΩ, 10-kΩ potentiometer.
☐ Capacitors: 0.01-μF

EQUIPMENT USED

Instrument	Manufacturer/Model No.	Serial No.
DMM		
Dual power supply		
Oscilloscope		
Signal Generator		

Table 34-1

TEXT REFERENCE

Circuit Analysis with Devices: Theory and Practice
Section 31.3   INTEGRATORS AND DIFFERENTIATORS
Section 31.4   INSTRUMENTATION AMPLIFIERS

MEASUREMENTS

1. Construct the circuit shown in Figure 34-1.
2. Apply a sinusoidal input signal of v_{in} = 0.1 V_{p-p}. Measure and record on the following page, the waveforms observed at v_{out1}, v_{out2}, and v_{out}. The display of the input voltage is shown as a reference.

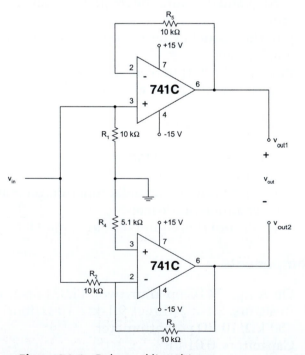

Figure 34-1   Balanced line driver

3. Calculate $A_{v1} = \dfrac{V_{out1}}{V_{in}}$, $A_{v2} = \dfrac{V_{out2}}{V_{in}}$, and $A_v = \dfrac{V_{out}}{V_{in}}$. Record the values adjacent to each display.

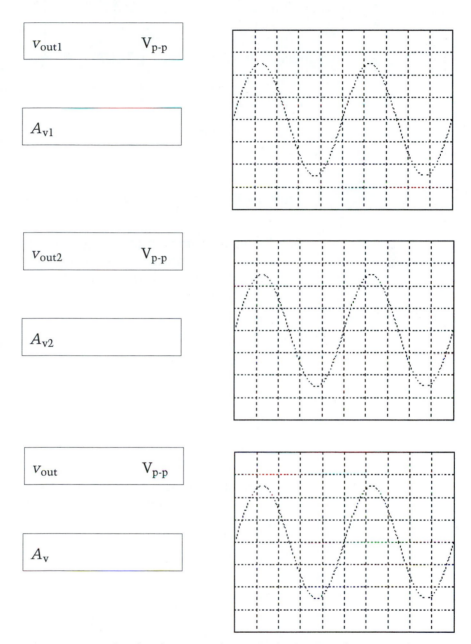

V_{out1}	$V_{p\text{-}p}$

A_{v1}	

V_{out2}	$V_{p\text{-}p}$

A_{v2}	

V_{out}	$V_{p\text{-}p}$

A_v	

4. Connect a 1-kΩ load across the output terminals. Does the output voltage change? (Circle the correct answer.)

yes/no

5. **Do not disassemble the circuit of Figure 34-1.** Construct the circuit shown in Figure 34-2. Use the TL071 op-amp. The pinout for this op-amp is identical to that of the 741C.

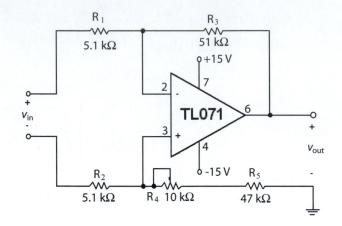

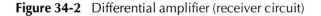

Figure 34-2 Differential amplifier (receiver circuit)

6. Connect a common mode signal of approximately 10 $V_{p\text{-}p}$ to the input of the amplifier. Adjust the 10-kΩ potentiometer to minimize the output voltage. Measure the common-mode output voltage and calculate the common-mode gain of the amplifier.

$v_{in(cm)}$	$V_{p\text{-}p}$
$v_{out(cm)}$	$V_{p\text{-}p}$
A_{cm}	

7. Connect the input of the receiver circuit (Figure 34-2) to the output of the line driver (Figure 34-1). Measure the overall voltage gain of the resulting circuit.

$v_{in(driver)}$	$V_{p\text{-}p}$
$v_{out(receiver)}$	$V_{p\text{-}p}$
A_{total}	

8. Calculate the differential voltage, A_d, of the receiver section.

A_d	

9. Calculate the common-mode rejection ratio.

$$\text{CMRR} = \left| \frac{A_{\text{d}}}{A_{\text{cm}}} \right| \qquad (34\text{-}1)$$

CMRR

10. Calculate the common-mode rejection ratio in decibels.

$$[\text{CMMR}]_{\text{dB}} = 20 \log \text{CMRR} \qquad (34\text{-}2)$$

CMRR	dB

11. Construct the circuit shown in Figure 34-3.
12. Apply a sinusoidal input signal of $v_{\text{in}} = 0.1\ V_{\text{p-p}}$. Measure and record the output voltage, v_{out} at the frequencies shown in Table 34-1. Calculate voltage gain.

Frequency	$v_{\text{out}}\ (V_{\text{p-p}})$	A_{v}	$[A_{\text{v}}]_{\text{dB}}$
10 Hz			
20 Hz			
50 Hz			
100 Hz			
200 Hz			
1 kHz			
2 kHz			
5 kHz			
10 kHz			
20 kHz			
50 kHz			
100 kHz			

Table 34-1 Frequency response measurements

13. Plot the data of Table 34-1 as a semi-logarithmic graph (Graph 34-1).

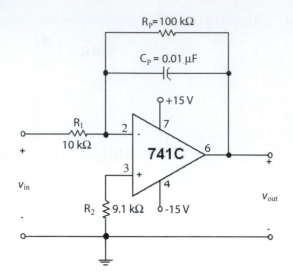

Figure 34-3 Integrator

14. Apply a 10-kHz square wave input having v_{in} = 5.0 V_{p-p}. Measure the output with an ac-coupled oscilloscope and record your observation in the space provided. Indicate the correct peak-to-peak output voltage and phase relationship.

voltage gain, dB

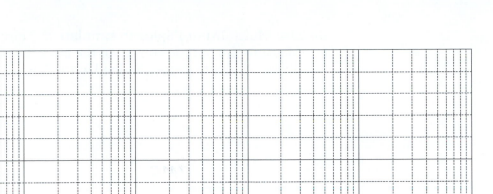
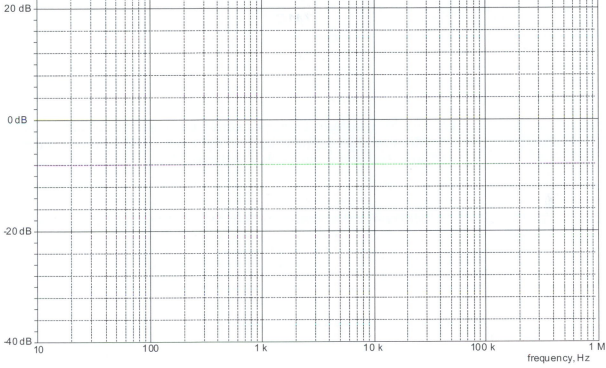

Graph 34-1   Frequency response of an op-amp integrator

FOR FURTHER INVESTIGATION AND DISCUSSION

15. Use calculus to derive the expression for output voltage for the circuit of Figure 34-3. Compare the theoretical calculations to the actual measurements.

16. Use MultiSIM or PSpice to simulate the circuit of Figure 34-1. Measure the output voltage of the circuit and compare the predicted results to those obtained in the lab.

17. Use MultiSIM or PSpice to simulate the circuit of Figure 34-3. Measure the output voltage of the circuit and compare the predicted results to those obtained in the lab.

Op-Amp Characteristics

Objectives

After completing this lab, you will be able to

- use the NULL OFFSET terminals of an op-amp to minimize the output offset voltage,
- determine the slew rate of an op-amp from the manufacturer's specifications,
- given the slew rate, determine the maximum sinusoidal frequency that can be applied to an op-amp for a given amplitude,
- given the slew rate, determine the maximum amplitude of a sinusoidal waveform for a given frequency,
- measure the slew rate of an op-amp using a square-wave input.

Equipment Required

- ☐ Digital multimeter (DMM)
- ☐ Dual power supply
- ☐ Signal generator (sinusoidal function generator)
- ☐ Two-channel oscilloscope
  Note: Record this equipment in Table 35-1.

Components

- ☐ Op-Amps: 741C op-amp
- ☐ Resistors: 1-kΩ, 10-kΩ (four), 10-kΩ potentiometer.

EQUIPMENT USED

Instrument	Manufacturer/Model No.	Serial No.
DMM		
Dual power supply		
Oscilloscope		
Signal Generator		

Table 35-1

TEXT REFERENCE

Circuit Analysis with Devices: Theory and Practice
Section 30.6   OP-AMP SPECIFICATIONS

MEASUREMENTS

1. Construct the circuit shown in Figure 35-1.

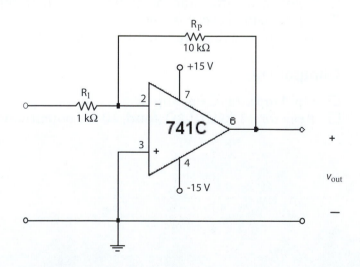

Figure 35-1   Offset measurement circuit

2. With no voltage applied at the input, measure the output voltage with a sensitive DMM. Record the offset voltage.

$V_{out(offset)}$	V

3. Place a 10-kΩ potentiometer between the NULL-OFFSET terminals of the op-amp as shown in Figure 35-2. Adjust the potentiometer so that $V_{out(offset)} = 0$ V.
4. Refer to the manufacturer's specification of the 741 op-amp and determine the slew rate. Use the correct units for this specification and record the value in the space provided below.

Slew rate	

5. Apply an input sinusoidal input signal so that the output voltage of the op-amp is $v_{out} = 20$ V_{p-p}. Adjust the frequency of the generator until the output becomes distorted (appears triangular). For a 20 V_{p-p} (10-V amplitude) at approximately what maximum frequency is the output no longer distorted?

f_{max}	Hz

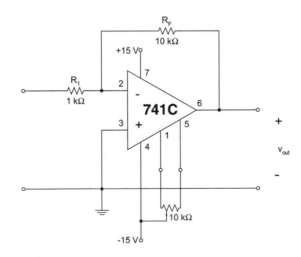

Figure 35-2 NULL-OFFSET adjust

6. Calculate the theoretical maximum frequency from the following expression:

$$2\pi f A \leq \text{slew rate} \qquad (35\text{-}1)$$

$f_{\text{max(theor)}}$	Hz

7. Apply an input sinusoidal input signal at a frequency of 20 kHz. While observing the output of the op-amp, adjust the amplitude of the sine wave until the output just begins to distort. What is the maximum peak-to-peak voltage at which the output is not distorted? What is the amplitude?

$V_{\text{out(max)}}$	$V_{\text{p-p}}$

$V_{\text{out(max)}}$	$V_{\text{p-p}}$

8. Use Equation 35-1 to calculate the theoretical maximum amplitude.

$V_{\text{out(max)}}$	$V_{\text{p-p}}$

9. Adjust the signal generator to provide a 1-kHz square wave with an amplitude of 10 V_{p} (20 $V_{\text{p-p}}$). Measure and record the slew rate of the waveform that you observe at the output of the op-amp. You may need to use the $\times$ 10 magnifier on the oscilloscope to determine the time.

Slew rate

NAME _____

DATE _____

CLASS _____

Schmitt Trigger and Free-running Multivibrator

Objectives

After completing this lab, you will be able to
- construct an op-amp Schmitt trigger circuit,
- observe and measure the voltage waveforms at the input and output of a Schmitt trigger circuit,
- predict the UTP (upper trip point) and LTP (lower trip point) of a Schmitt trigger circuit,
- modify the Schmitt trigger circuit as a free-running multivibrator (square-wave generator).

Equipment Required

☐ Digital multimeter (DMM)
☐ Dual power supply
☐ Signal generator (sinusoidal function generator)
☐ Two-channel oscilloscope
   Note: Record this equipment in Table 36-1.

Components

☐ Op-Amps: 741C op-amp
☐ Resistors: 1-kΩ, 5.1-kΩ, 10-kΩ, 12-kΩ
☐ Capacitors: 0.47 µF radial lead film.

EQUIPMENT USED

Instrument	Manufacturer/Model No.	Serial No.
DMM		
Dual power supply		
Oscilloscope		
Signal Generator		

Table 36-1

TEXT REFERENCE

Circuit Analysis with Devices: Theory and Practice
Section 32.1   BASICS OF FEEDBACK

MEASUREMENTS

1. Construct the circuit shown in Figure 36-1.

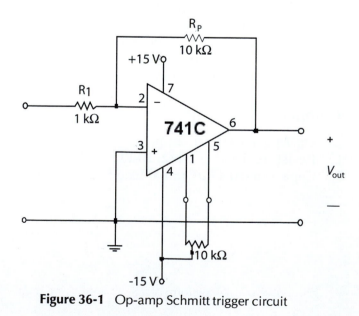

Figure 36-1   Op-amp Schmitt trigger circuit

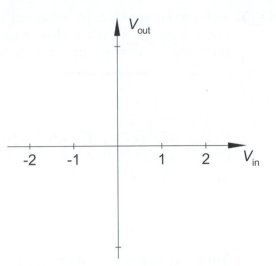

Figure 36-2 Transfer curve for a Schmitt trigger

2. Assuming that V_{SAT} = 13 V, sketch the transfer curve for the circuit (Figure 36-2).
3. Apply a triangular waveform having a frequency f = 100 Hz to the input of the circuit of Figure 36-1. Let v_{in} = 4 $V_{p\text{-}p}$. Simultaneously display both the input and output voltages on the oscilloscope.
4. In the space provided in figure 36-3, sketch the output voltage, showing all voltage and time measurements. Indicate the vertical setting used for Channel 2.

Horizontal:	1 ms/Div

Channel 1 Vertical:	1 V/Div

Channel 2 Vertical:	V/Div

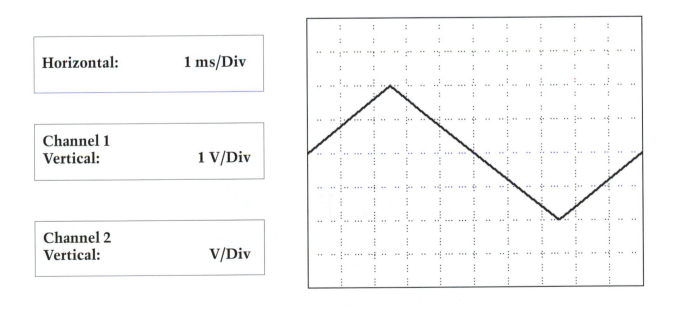

Figure 36-3 Output waveform of a Schmitt trigger circuit

5. Refer to Figure 36-2. For what value of input voltage does the output voltage go from a high voltage to a low voltage? This represents the upper trip point (UTP) of the Schmitt trigger.

UTP	V

6. Refer to Figure 36-2. For what value of input voltage does the output voltage go from a low voltage to a high voltage? This represents the lower trip point (LTP) of the Schmitt trigger

LTP	V

7. Turn off the power to the voltage source and modify the previous circuit as shown in Figure 36-4.
8. Measure and record the voltages observed across the capacitor and the output voltage observed in the circuit of Figure 36-4. Sketch the output observed in the space provided on the following page. Label each display and provide the oscilloscope settings in the space provided.

f	Hz

v_{out}	$V_{p\text{-}p}$

v_C	$V_{p\text{-}p}$

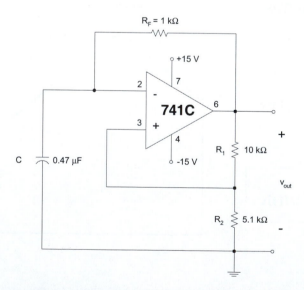

Figure 36-4 Free-running multivibrator circuit

<table>
<tr><td>Horizontal:</td><td>s/Div</td></tr>
</table>

<table>
<tr><td>Channel 1
Vertical:</td><td>V/Div</td></tr>
</table>

<table>
<tr><td>Channel 2
Vertical:</td><td>V/Div</td></tr>
</table>

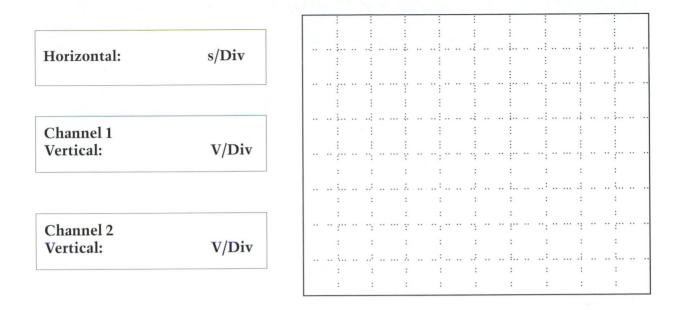

FOR FURTHER INVESTIGATION AND DISCUSSION

9. Assuming that the output voltage of the free-running multivibrator circuit of Figure 36-4 starts at $V_{out} = +V_{SAT} = +13$ V, describe the operation of the circuit during one full period of operation. As part of your discussion, include calculations involving the appropriate charge and discharge times and capacitor voltages. Determine the theoretical period of oscillation and compare this result to the actual measured period.

10. Use either MultiSIM or PSpice to obtain the period and frequency of oscillation of the multivibrator circuit. Compare this result to the actual measured value and offer an explanation for any discrepancy.

Op-Amp Function Generator

Objectives

After completing this lab, you will be able to
- construct an op-amp multivibrator, integrator, and filter to produce a function generator,
- observe and measure the voltage waveforms at the input and output of the various parts of the function generator,
- calculate the predicted results from theoretical values,
- use MultiSIM to observe the predicted results.

Equipment Required

☐ Digital multimeter (DMM)
☐ Dual power supply
☐ Signal generator (sinusoidal function generator)
☐ Two-channel oscilloscope
  Note: Record this equipment in Table 37-1.

Components

☐ Op-Amps: 741C op-amp (two)
☐ Resistors: 1-kΩ (three), 10-kΩ, 12-kΩ, 100-kΩ
☐ Capacitors: 0.1-μF, 0.47 μF radial lead film.

EQUIPMENT USED

Instrument	Manufacturer/Model No.	Serial No.
DMM		
Dual power supply		
Oscilloscope		
Signal Generator		

Table 37-1

TEXT REFERENCE

Circuit Analysis with Devices: Theory and Practice
Section 32.1   BASICS OF FEEDBACK
Section 32.2   RELAXATION OSCILLATOR

MEASUREMENTS

1. Construct the circuit shown in Figure 37-1.

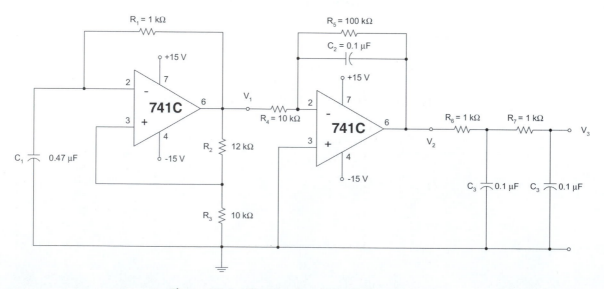

Figure 37-1   Op-amp function generator circuit

2. Sketch the voltage appearing at the points labeled as V_1 and V_2 showing all oscilloscope settings in the space provided below.

Horizontal: _____ s/Div

Channel 1
Vertical: _____ V/Div

Channel 2
Vertical: _____ V/Div

3. Sketch the voltage appearing at the points labeled as V_1 and V_3 showing all oscilloscope settings in the space provided below.

Horizontal: _____ s/Div

Channel 1
Vertical: _____ V/Div

Channel 2
Vertical: _____ V/Div

4. Determine the frequency of oscillation from your measurements.

f_{measured}	Hz

5. Using the component values provided, calculate the theoretical frequency of oscillation.

$f_{\text{theorectical}}$	Hz

6. Determine the cutoff frequency for the filter section.

f_{cutoff}	Hz

FOR FURTHER INVESTIGATION AND DISCUSSION

7. Use either MultiSIM or PSpice to simulate the circuit of Figure 37-1. Obtain a printout of the oscilloscope displays (MultiSIM) or the Probe postprocessor (PSpice) and submit these with the lab.

NAME _____

DATE _____

CLASS _____

Phase-shift Oscillator

Objectives

After completing this lab, you will be able to
- construct an op-amp Schmitt trigger circuit,
- construct a phase-shift oscillator,
- explain the general operation of the phase-shift oscillator,
- measure the output frequency of a phase-shift oscillator and compare the results to the theoretical value,
- use zener diodes to improve the operation of a phase-shift oscillator).

Equipment Required

☐ Digital multimeter (DMM)
☐ Dual power supply
☐ Signal generator (sinusoidal function generator)
☐ Two-channel oscilloscope
Note: Record this equipment in Table 38-1.

Components

☐ Op-Amps: 741C op-amp
☐ Resistors: 1-kΩ (three), 22-kΩ, 27-kΩ, 20-kΩ potentiometer
☐ Capacitors: 0.022-μF radial lead film (three)
☐ Diodes: 1N4733A 5.1-V zener (two).

EQUIPMENT USED

Instrument	Manufacturer/Model No.	Serial No.
DMM		
Dual power supply		
Oscilloscope		
Signal Generator		

Table 38-1

TEXT REFERENCE

Circuit Analysis with Devices: Theory and Practice
Section 32.4   THE PHASE-SHIFT OSCILLATOR

DISCUSSION

1. Figure 38-1 shows the circuit of a typical phase-shift oscillator.

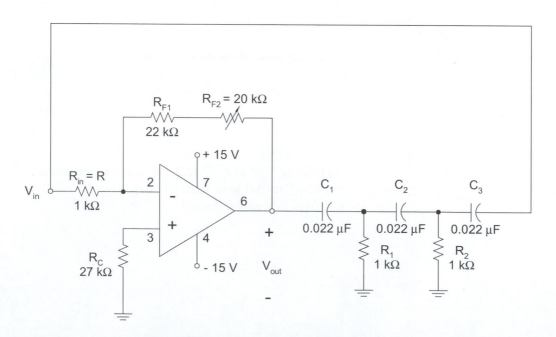

Figure 38-1   The phase-shift oscillator

In the circuit of Figure 38-1, the resistors R_1, R_2, and R_{in}, in combination with capacitors C_1, C_2, and C_3 set up the feedback path. In order for the amplifier to oscillate, the feedback path must provide for 180° of phase shift and the amplifier must have sufficient gain to provide for an overall gain of unity (1) for the circuit.

The resonant frequency of the phase-shift oscillator is determined from the following expression:

$$f = \frac{1}{2\pi\sqrt{6}\,RC} \qquad (38\text{-}1)$$

The R-C network has a gain of 1/29 at this frequency, and so the op-amp must provide for the balance of the gain. In other words, the gain of the op-amp must be $A_v = -29$ to result in an overall gain of unity (1).

2. Determine the theoretical resonant frequency of the oscillator.

f	Hz

MEASUREMENTS

3. Connect the circuit and adjust the feedback resistor, R_{F2} for a value of 20 kΩ. This value will provide sufficient gain for the amplifier to begin to oscillate. However, the gain will likely be too large, resulting in distortion. Sketch the resultant waveform in the space provided. Show the oscilloscope settings beside the display.

Horizontal:	s/Div

Channel 1 Vertical:	V/Div

Channel 2 Vertical:	V/Div

4. Adjust the potentiometer to result in minimum distortion. Measure the observed frequency.

5. What happens if you continue to lower the value of the potentiometer resistance?

_____

_____

_____

6. Readjust R_{F2} to provide for the least distortion, while maintaining the oscillation. Turn off the power to the circuit and then turn the power back on. Did oscillations resume?

_____

7. The operation of the phase-shift oscillator circuit can be improved by introducing a variable feedback path through two zener diodes as shown in Figure 38-2. When the voltage at the output is low, the zener diodes are effectively open circuits, allowing a large amount of gain in the op-amp circuit. As the voltage at the output increases, the zener diodes will maintain a constant voltage across feedback resistance.

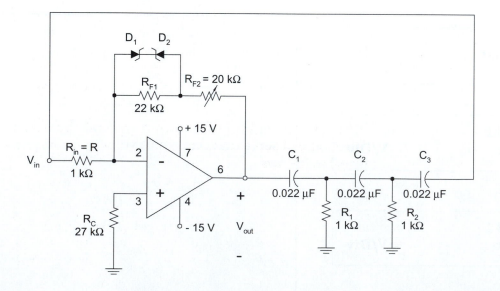

Figure 38-2   Phase-shift oscillator with zener diode distortion control

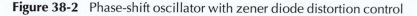

8. Modify the circuit as illustrated in Figure 38-2. Adjust R_{F2} to provide for the least distortion, while maintaining the oscillation. Turn off the power and then turn it back on. Did oscillations resume?

_____

9. Sketch the resultant waveform and indicate the observed frequency and amplitude. Show the oscilloscope settings beside the display.

Horizontal:	s/Div

Channel 1 Vertical:	V/Div

Channel 2 Vertical:	V/Div

f	Hz

Measured amplitude	$V_{p\text{-}p}$

10. The zener diodes control the amplitude of the output voltage. You should observe that the output voltage has amplitude determined as follows

$$V_p = 2V_z \qquad (38\text{-}2)$$

11. Use the above expression to calculate the theoretical amplitude.

Theoretical amplitude	V_{p-p}

12. Compare your measured amplitude to the predicted value.

_____

_____

_____

FOR FURTHER INVESTIGATION AND DISCUSSION

Use either MultiSIM or PSpice to simulate the operation of the circuit in Figure 38-1. Compare the observed frequency to the actual measured value and offer an explanation for any discrepancy.

NAME _____

DATE _____

CLASS _____

LAB 39

Applications of a 555 Timer

Objectives

- After completing this lab, you will be able to
- construct a relaxation oscillator using a 555 timer,
- explain the basic operation of a 555 timer,
- calculate the frequency of oscillation of a 555 timer,
- use a 555 timer to control the duration of an alarm.

Equipment Required

☐ dc power supply
☐ Two-channel oscilloscope
   Note: Record this equipment in Table 39-1.

Components

☐ ICs: NE555 timer (or equivalent)
☐ Resistors: 1.2-kΩ, 7.5-kΩ (two), 51-kΩ 1.0-MΩ
☐ Capacitors: 0.01-μF, 0.1-μF radial lead film, 4.7-μF electrolytic
☐ Diodes: Light emitting diode
☐ Miscellaneous: Momentary SPDT switch (optional)

EQUIPMENT USED

Instrument	Manufacturer/Model No.	Serial No.
dc power supply		
Oscilloscope		

Table 39-1

TEXT REFERENCE

Circuit Analysis with Devices: Theory and Practice
Section 32.7   THE 555 TIMER

DISCUSSION

1. Figure 39-1 shows the internal circuit of a 555 Timer.

    In the circuit shown in Figure 39-1, the 5-kΩ resistors form a voltage divider network. Each resistor will drop a voltage of $V_{CC}/3$. These voltages are then applied to the inputs of two comparators, with the outputs of the comparators applied to an R-S flip-flop circuit.

    If the *control* voltage is less than $V_{CC}/3$, then the output of the bottom comparator will go high, causing the R-S flip-flop to *set* ($\overline{Q} = 0$). The output of the 555 timer will be $+ V_{CC}$. The discharge transistor will be off.

    If the *threshold* voltage goes above $2 V_{CC}/3$, then the output of the top comparator will go high, causing the R-S flip-flop to *reset* ($\overline{Q} = 1$). The output of the 555 timer will be zero volts. The B-E junction of the discharge transistor will be forward biased.

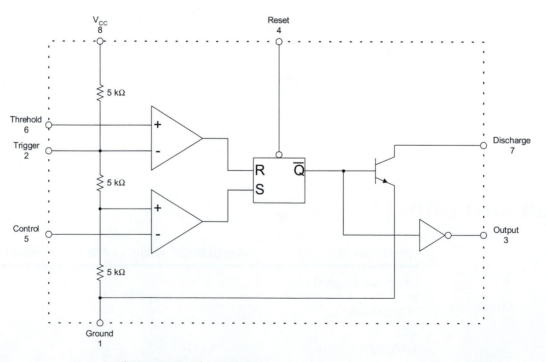

Figure 39-1 The 555 timer

When used as a relaxation oscillator as illustrated in Figure 39-2, the period of oscillation is determined as

$$T = (\ln 2)(R_A + 2R_B)C \tag{39-1}$$

which results in a frequency of oscillation of

$$f = \frac{1}{(\ln 2)(R_A + 2R_B)C} \approx \frac{1.44}{(R_A + 2R_B)C} \tag{39-2}$$

CALCULATIONS

2. Determine the theoretical frequency of the circuit shown in Figure 39-2.

$f_{\text{theoretical}}$		Hz

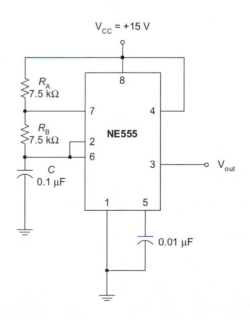

$V_{CC} = +15$ V

Figure 39-2   Relaxation oscillator

MEASUREMENTS

3. Build the circuit as shown in Figure 39-2.
4. Measure and record the actual frequency of operation.

$f_{measured}$ Hz

5. Record the waveform observed across the capacitor. Sketch the resultant waveform in the space provided. Show the oscilloscope settings beside the display.

Horizontal: s/Div

Channel 1
Vertical: V/Div

6. Record the waveform observed at the output. Sketch the resultant waveform in the space provided. Show the oscilloscope settings beside the display.

Horizontal:	s/Div

Channel 1 Vertical:	V/Div

DISCUSSION

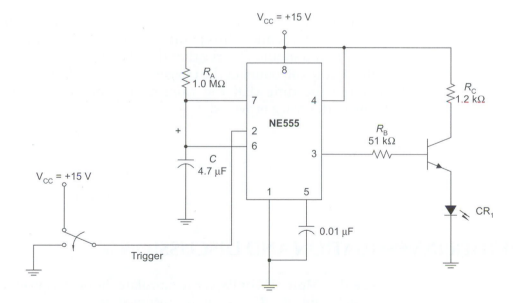

Figure39-3

7. Besides providing astable operation as a relaxation oscillator, the 555 timer can be used as a monostable circuit, providing a single pulse each time the trigger voltage goes low. The resulting output

pulse will have a minimum duration, regardless of how short the trigger pulse was. This condition is ideally suited for alarm circuits, which will continue to provide an alarm, even if the alarm trigger was reset. The duration of the output pulse is determined by the timing circuit, and can be adjusted to be active from several seconds up to many minutes. The monostable circuit is shown in Figure 39-3.

The duration of the pulse is determined as

$$T_w = (\ln 3)R_A C \approx 1.10 R_A C \qquad\qquad (39\text{-}3)$$

CALCULATIONS

8. Determine the theoretical duration of the pulse that will resulting when the trigger of the 555 timer in Figure 39-3 goes low.

$$\boxed{T_{\text{theoretical}} \qquad\qquad\qquad \text{s}}$$

MEASUREMENTS

9. Use a momentary switch or manually remove the trigger from V_{CC} = 15 V and connect this point to ground as illustrated in Figure 39-3. You should observe that the LED representing an alarm will come on. Reconnect the trigger to V_{CC} = 15 V. Use a watch to measure the time that the "alarm" remains on. Record this value in the space provided.

$$\boxed{T_{\text{measured}} \qquad\qquad\qquad \text{s}}$$

FOR FURTHER INVESTIGATION AND DISCUSSION

10. Use either MultiSIM or PSpice to simulate the operation of the circuit in Figure 39-2. Compare the observed frequency to the actual measured value and offer an explanation for any discrepancy.
11. Describe the operation of the monostable circuit shown in Figure 39-3. In your discussion, include the expected capacitor voltage and show how the pulse duration is determined by the charging of the capacitor.

NAME _____

DATE _____

CLASS _____

Voltage Controlled Oscillator

Objectives

After completing this lab, you will be able to
- explain the basic principles of operation of the LM566 Voltage Controlled Oscillator,
- calculate the frequency of oscillation of a VCO circuit,
- determine the minimum modulating voltage of a VCO circuit,
- measure the output voltages of a VCO circuit,
- sketch the relationship of frequency as a function of the modulating voltage of a VCO,
- apply a sinusoidal input signal at the input of a VCO and observe the frequency modulated output voltage.

Equipment Required

☐ dc power supply
☐ Digital multimeter
☐ Signal generator
☐ Two-channel oscilloscope
  Note: Record this equipment in Table 40-1.

Components

☐ ICs: LM566C voltage controlled oscillator
☐ Resistors: 510-Ω, 10-kΩ, 5-kΩ potentiometer, 18-kΩ, 1.0-MΩ
☐ Capacitors: 4.7-nF radial lead film, 10-µF electrolytic

EQUIPMENT USED

Instrument	Manufacturer/Model No.	Serial No.
DMM		
Dual power supply		
Oscilloscope		
Signal Generator		

Table 40-1

TEXT REFERENCE

Circuit Analysis with Devices: Theory and Practice
Section 32.8   THE VOLTAGE CONTROLLED OSCILLATOR - VCO

DISCUSSION

1. Figure 40-1 shows the internal circuit of the LM566C voltage controlled oscillator.

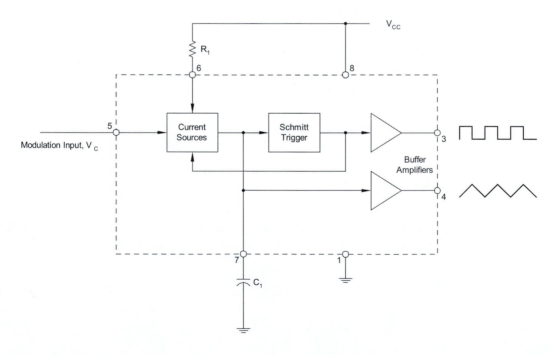

Figure 40-1   The LM566C voltage controlled oscillator

2. The voltage controlled oscillator, VCO, is a device that generates both a square wave and a triangular wave. A timing resistor, R_1, a timing capacitor, C_1, and an externally applied modulating voltage, V_C, determine the frequency of oscillation as follows:

$$f_0 = \frac{2.4}{R_1 C_1}\left(\frac{V_{CC} - V_C}{V_{CC}}\right) \qquad (40\text{-}1)$$

CALCULATIONS

3. Refer to the manufacturer's specifications to answer the following questions:
   a. What is the maximum operating frequency of the LM566C VCO?

f_{max}	Hz

   b. If the supply voltage is selected to be $V_{CC} = 15$ V, what is the minimum voltage that must be applied to the modulation input, V_C?

$V_{C(max)}$	V

4. Refer to the circuit shown in Figure 40-2. What is the minimum voltage that you expect at the modulating input terminal?

$V_{C(min)}$	V

5. What is the maximum voltage that you expect at the modulating input terminal?

$V_{C(max)}$	V

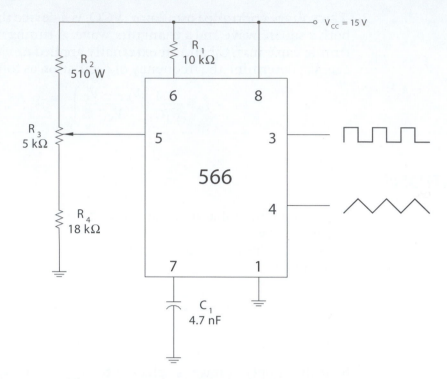

Figure 40-2   Voltage controlled oscillator circuit

6. Using the values from Steps 4 and 5, determine the frequency range of the output signal as the potentiometer is varied from 0 to 5 kΩ.

$f =$ _____ Hz to _____ Hz

MEASUREMENTS

7. Construct the circuit of Figure 40-2. Adjust the 5-kΩ potentiometer so that V_C is at its minimum value. Record this value.

$V_{C(min)}$	V

8. Place your oscilloscope into its dc mode. Measure and record the signal appearing at each of the outputs. Show the oscilloscope settings beside the display. Label the reference voltage (0 V) for each waveform.

Square Wave Output:

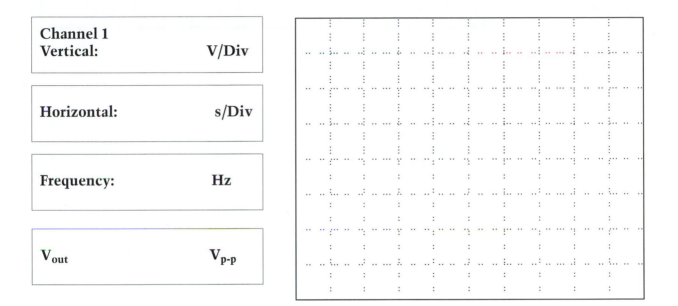

Channel 1
Vertical: **V/Div**

Horizontal: **s/Div**

Frequency: **Hz**

V_{out} $V_{p\text{-}p}$

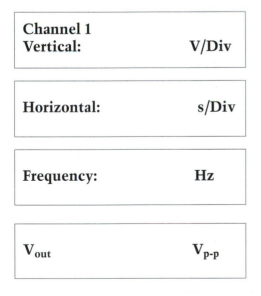

Channel 1
Vertical: **V/Div**

Horizontal: **s/Div**

Frequency: **Hz**

V_{out} $V_{p\text{-}p}$

Sawtooth Wave Output:

9. Adjust the 5-kΩ potentiometer so that V_C has the quantities indicated in Table 40-1. Measure and record the corresponding amplitude and frequency of the square wave for each voltage setting.

V_C	f	$V_{out(p\text{-}p)}$
11.5 V		
12.0 V		
12.5 V		
13.0 V		
13.5 V		
14.0 V		

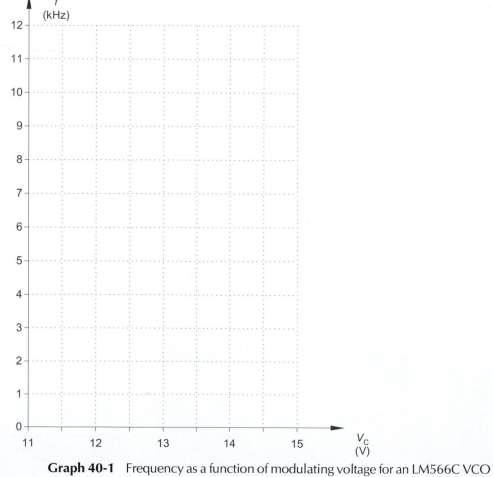

Graph 40-1   Frequency as a function of modulating voltage for an LM566C VCO

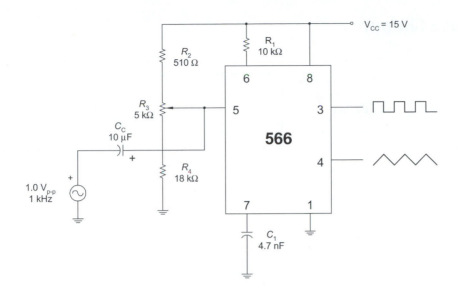

Figure 40-3 Voltage controlled oscillator used as a frequency modulator circuit

Table 40-1 Amplitude and frequency of the square wave as a function of modulating voltage

10. Plot the frequency as a function of modulating voltage in the Graph 40-1. Join the points using the best smooth continuous curve.
11. Modify the VCO circuit as shown in Figure 40-3.
12. Set the signal generator to provide a 1-V_{p-p} sinusoidal waveform at a frequency of 1 kHz. Connect the sinusoidal voltage to Ch. 1 of your oscilloscope. Use Ch. 2 of your oscilloscope to observe the resulting square wave output signal of the VCO circuit. Sketch the result in the space provided on the following page. In order to observe the FM signal output signal, you will need to adjust your oscilloscope to provide a single trace.

Square Wave Output:

Channel 1
Vertical: **V/Div**

Horizontal: **s/Div**

Frequency: **Hz**

V_{out} $V_{p\text{-}p}$

<table>
<tr><td>LAB
41</td></tr>
</table>

SCR Triggering Circuit

Objectives

After completing this lab, you will be able to
- build a test circuit to determine the *peak point*, *valley point*, and the *intrinsic standoff ratio* of a unijunction transistor (UJT),
- build a UJT relaxation oscillator and compare the predicted frequency to the actual measured frequency,
- build a test circuit to examine the operation of a silicon controlled rectifier (SCR) and determine the approximate *holding current* of the SCR.

Equipment Required

☐ dc power supply
☐ Digital multimeter
☐ Two-channel oscilloscope
Note: Record this equipment in Table 41-1.

Components

☐ Resistors: 47-Ω, 120-Ω, 1-kΩ (two), 20-kΩ potentiometer
☐ Capacitors: 0.33-μF radial lead film.
☐ Miscellaneous: C106 Silicon Controlled Rectifier, 2N4870 Unijunction Transistor, 24-V transformer (Ensure that the primary side of the transformer is safely connected to 120 VAC.)

EQUIPMENT USED

Instrument	Manufacturer/Model No.	Serial No.
dc power supply		
Signal generator		
Oscilloscope		

Table 41-1

TEXT REFERENCE

Circuit Analysis with Devices: Theory and Practice
Section 33.2   TRIGGER DEVICES
Section 33.3   SILICON CONTROLLED RECTIFIERS - SCRS

MEASUREMENTS

1. Obtain the manufacturer's specification sheet for the 2N4870 (UJT) Unijunction Transistor. Determine the location of the base terminals, B_1 and B_2. Using a DMM ohmmeter, measure and record the interbase resistance, R_{BB}.

R_{BB}	Ω

2. Build the UJT test circuit illustrated in Figure 41-1.

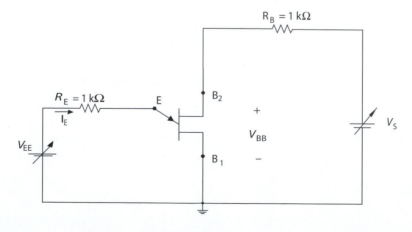

Figure 41-1   UJT test circuit

3. Adjust V_S to provide a voltage of $V_S = 10.0$ V.
4. With a DMM voltmeter connected between the emitter and ground, slowly increase V_{BB} until V_E suddenly drops. The maximum value to which V_E rises is called the *peak point*, and is designated as V_P. Record this value.

V_P	V

5. Increase V_{EE} and observe that V_E also increases. Reduce V_{EE} and observe that V_E once again increases. You have located the *valley point* of the UJT characteristic curve. For the valley point, measure and record the valley voltage, V_V and the valley current, I_V. Note that you "measure" the current by measuring the voltage across the 1-kΩ emitter resistance and then use Ohm's law to calculate the current.

V_V	V

I_V	mA

6. Determine the *intrinsic standoff ratio*, η by applying the following equation:

$$V_p = \eta v V_{BB} + V_D \qquad (41\text{-}1)$$

where V_D is the voltage across a forward-biased silicon diode. Let $V_D = 0.6$ V.

η	

7. Refer to the manufacturer's specifications and provide typical values for and R_{BB}.

η	

R_{BB}	kΩ

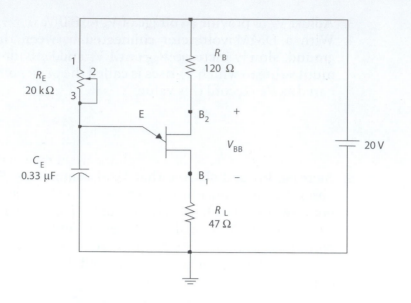

Figure 41-2   UJT relaxation oscillator

8. Construct the UJT relaxation oscillator circuit shown in Figure 41-2. Connect Channel 1 of the oscilloscope between the emitter and ground. Connect Channel 2 between base B1 and ground.
9. Vary the 20-kΩ potentiometer and observe the effect. Adjust the potentiometer to result in an oscillator frequency of 500 Hz. Sketch the waveforms observed at the emitter and at B1. Label each measurement and show the oscilloscope settings beside the display.

Horizontal:	s/Div

Channel 1:	
Vertical:	V/Div

Channel 2:	
Vertical:	V/Div

10. Disconnect the supply voltage and carefully remove the potentiometer from the circuit without changing its value. Measure and record the resistance between terminals 1 and 2 of the potentiometer.

R_E Ω

11. Calculate the theoretical oscillator frequency using the following expression. Compare the theoretical frequency to the actual measured frequency.

$$f \approx \frac{1}{R_E C_E \ln\left(\frac{1}{1-\eta}\right)}$$

(41-2)

f Hz

12. Modify the circuit as illustrated in Figure 41-3.

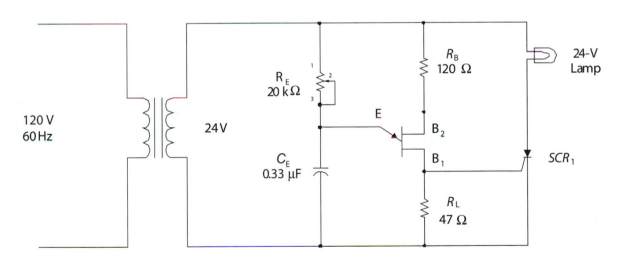

Figure 41-3   UJT phase control of an SCR

13. Adjust the potentiometer to provide a firing angle of 90°. Use the oscilloscope to observe the voltage across the load. Sketch the waveform and show the oscilloscope settings beside the display.

Horizontal: s/Div

Channel 1:
Vertical: V/Div

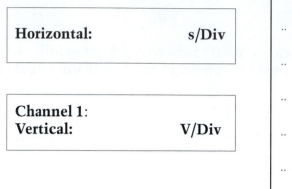

14. Use the oscilloscope to observe the voltage across the SCR. Sketch the waveform and show the oscilloscope settings beside the display.

Horizontal: s/Div

Channel 1:
Vertical: V/Div

Appendix

A

Resistor Color Codes

Figure A-1 shows a typical color coded resistor together with the standard colors and corresponding designations. Table A-1 gives nominal resistor values for resistors having 5%, 10%, and 20% values.

First Band — 1st Digit

Second Band — 2nd Digit

Color	Digit	Multiplier	Tolerance	Reliability Level (Percent Per 1000 Hours) (When Applicable)
Black	0	1	—	—
Brown	1	10	—	M = 1.0%
Red	2	100	—	P = 0.1%
Orange	3	1000	—	R = 0.01%
Yellow	4	10,000	—	S = 0.001%
Green	5	100,000	—	—
Blue	6	1,000,000	—	—
Violet	7	10,000,000	—	—
Gray	8	—	—	—
White	9	—	—	—
Gold	—	0.1	± 5%	—
Silver	—	—	± 10%	—
No color	—	—	± 20%	—

Figure A-1   Standard resistor values (Courtesy of the Allen-Bradley Company)

Resistance Color Code / Nominal Resistance in Ohms — TOLERANCE COLOR CODE

1st BAND (1st digit)	2nd BAND (2nd digit)	3rd BAND (Number of zeros after 1st and 2nd digit)	4th BAND Gold ±5%	Silver ±10%	None ±20%
Brown	Black	Gold	1.0	1.0	1.0
Brown	Brown	Gold	1.1	—	—
Brown	Red	Gold	1.2	1.2	—
Brown	Orange	Gold	1.3	—	—
Brown	Green	Gold	1.5	1.5	1.5
Brown	Blue	Gold	1.6	—	—
Brown	Gray	Gold	1.8	1.8	—
Red	Black	Gold	2.0	—	—
Red	Red	Gold	2.2	2.2	2.2
Red	Yellow	Gold	2.4	—	—
Red	Violet	Gold	2.7	2.7	—
Orange	Black	Gold	3.0	—	—
Orange	Orange	Gold	3.3	3.3	3.3
Orange	Blue	Gold	3.6	—	—
Orange	White	Gold	3.9	3.9	—
Yellow	Orange	Gold	4.3	—	—
Yellow	Violet	Gold	4.7	4.7	4.7
Green	Brown	Gold	5.1	—	—
Green	Blue	Gold	5.6	5.6	—
Blue	Red	Gold	6.2	—	—
Blue	Gray	Gold	6.8	6.8	6.8
Violet	Green	Gold	7.5	—	—
Gray	Red	Gold	8.2	8.2	—
White	Brown	Gold	9.1	—	—
Brown	Black	Black	10	10	10
Brown	Brown	Black	11	—	—
Brown	Red	Black	12	12	—
Brown	Orange	Black	13	—	—
Brown	Green	Black	15	15	15
Brown	Blue	Black	16	—	—
Brown	Gray	Black	18	18	—
Red	Black	Black	20	—	—
Red	Red	Black	22	22	22
Red	Yellow	Black	24	—	—
Red	Violet	Black	27	27	—
Orange	Black	Black	30	—	—
Orange	Orange	Black	33	33	33
Orange	Blue	Black	36	—	—
Orange	White	Black	39	39	—
Yellow	Orange	Black	43	—	—
Yellow	Violet	Black	47	47	47
Green	Brown	Black	51	—	—
Green	Blue	Black	56	56	—
Blue	Red	Black	62	—	—
Blue	Gray	Black	68	68	68
Violet	Green	Black	75	—	—
Gray	Red	Black	82	82	—
White	Brown	Black	91	—	—
Brown	Black	Brown	100	100	100
Brown	Brown	Brown	110	—	—
Brown	Red	Brown	120	120	—
Brown	Orange	Brown	130	—	—
Brown	Green	Brown	150	150	150
Brown	Blue	Brown	160	—	—
Brown	Gray	Brown	180	180	—
Red	Black	Brown	200	—	—
Red	Red	Brown	220	220	220
Red	Yellow	Brown	240	—	—
Red	Violet	Brown	270	270	—
Orange	Black	Brown	300	—	—
Orange	Orange	Brown	330	330	330
Orange	Blue	Brown	360	—	—
Orange	White	Brown	390	390	—
Yellow	Orange	Brown	430	—	—
Yellow	Violet	Brown	470	470	470

Resistance Color Code / Nominal Resistance in Ohms — TOLERANCE COLOR CODE

1st BAND (1st digit)	2nd BAND (2nd digit)	3rd BAND (Number of zeros after 1st and 2nd digit)	4th BAND Gold ±5%	Silver ±10%	None ±20%
Green	Brown	Brown	510	—	—
Green	Blue	Brown	560	560	—
Blue	Red	Brown	620	—	—
Blue	Gray	Brown	680	680	680
Violet	Green	Brown	750	—	—
Gray	Red	Brown	820	820	—
White	Brown	Brown	910	—	—
Brown	Black	Red	1000	1000	1000
Brown	Brown	Red	1100	—	—
Brown	Red	Red	1200	1200	—
Brown	Orange	Red	1300	—	—
Brown	Green	Red	1500	1500	1500
Brown	Blue	Red	1600	—	—
Brown	Gray	Red	1800	1800	—
Red	Black	Red	2000	—	—
Red	Red	Red	2200	2200	2200
Red	Yellow	Red	2400	—	—
Red	Violet	Red	2700	2700	—
Orange	Black	Red	3000	—	—
Orange	Orange	Red	3300	3300	3300
Orange	Blue	Red	3600	—	—
Orange	White	Red	3900	3900	—
Yellow	Orange	Red	4300	—	—
Yellow	Violet	Red	4700	4700	4700
Green	Brown	Red	5100	—	—
Green	Blue	Red	5600	5600	—
Blue	Red	Red	6200	—	—
Blue	Gray	Red	6800	6800	6800
Violet	Green	Red	7500	—	—
Gray	Red	Red	8200	8200	—
White	Brown	Red	9100	—	—
Brown	Black	Orange	10000	10000	10000
Brown	Brown	Orange	11000	—	—
Brown	Red	Orange	12000	12000	—
Brown	Orange	Orange	13000	—	—
Brown	Green	Orange	15000	15000	15000
Brown	Blue	Orange	16000	—	—
Brown	Gray	Orange	18000	18000	—
Red	Black	Orange	20000	—	—
Red	Red	Orange	22000	22000	22000
Red	Yellow	Orange	24000	—	—
Red	Violet	Orange	27000	27000	—
Orange	Black	Orange	30000	—	—
Orange	Orange	Orange	33000	33000	33000
Orange	Blue	Orange	36000	—	—
Orange	White	Orange	39000	39000	—
Yellow	Orange	Orange	43000	—	—
Yellow	Violet	Orange	47000	47000	47000
Green	Brown	Orange	51000	—	—
Green	Blue	Orange	56000	56000	—
Blue	Red	Orange	62000	—	—
Blue	Gray	Orange	68000	68000	68000
Violet	Green	Orange	75000	—	—
Gray	Red	Orange	82000	82000	—
White	Brown	Orange	91000	—	—

Nominal Resistance in Megohms

1st BAND	2nd BAND	3rd BAND	Gold ±5%	Silver ±10%	None ±20%
Brown	Black	Yellow	0.1	0.1	0.1
Brown	Brown	Yellow	0.11	—	—
Brown	Red	Yellow	0.12	0.12	—
Brown	Orange	Yellow	0.13	—	—
Brown	Green	Yellow	0.15	0.15	0.15
Brown	Blue	Yellow	0.16	—	—
Brown	Gray	Yellow	0.18	0.18	—
Red	Black	Yellow	0.20	—	—
Red	Red	Yellow	0.22	0.22	0.22
Red	Yellow	Yellow	0.24	—	—

Resistance Color Code / Nominal Resistance in Megohms — TOLERANCE COLOR CODE

1st BAND (1st digit)	2nd BAND (2nd digit)	3rd BAND (Number of zeros after 1st and 2nd digit)	4th BAND Gold ±5%	Silver ±10%	None ±20%
Red	Violet	Yellow	0.27	0.27	—
Orange	Black	Yellow	0.30	—	—
Orange	Orange	Yellow	0.33	0.33	0.33
Orange	Blue	Yellow	0.36	—	—
Orange	White	Yellow	0.39	0.39	—
Yellow	Orange	Yellow	0.43	—	—
Yellow	Violet	Yellow	0.47	0.47	0.47
Green	Brown	Yellow	0.51	—	—
Green	Blue	Yellow	0.56	0.56	—
Blue	Red	Yellow	0.62	—	—
Blue	Gray	Yellow	0.68	0.68	0.68
Violet	Green	Yellow	0.75	—	—
Gray	Red	Yellow	0.82	0.82	—
White	Brown	Yellow	0.91	—	—
Brown	Black	Green	1.0	1.0	1.0
Brown	Brown	Green	1.1	—	—
Brown	Red	Green	1.2	1.2	—
Brown	Orange	Green	1.3	—	—
Brown	Green	Green	1.5	1.5	1.5
Brown	Blue	Green	1.6	—	—
Brown	Gray	Green	1.8	1.8	—
Red	Black	Green	2.0	—	—
Red	Red	Green	2.2	2.2	2.2
Red	Yellow	Green	2.4	—	—
Red	Violet	Green	2.7	2.7	—
Orange	Black	Green	3.0	—	—
Orange	Orange	Green	3.3	3.3	3.3
Orange	Blue	Green	3.6	—	—
Orange	White	Green	3.9	3.9	—
Yellow	Orange	Green	4.3	—	—
Yellow	Violet	Green	4.7	4.7	4.7
Green	Brown	Green	5.1	—	—
Green	Blue	Green	5.6	5.6	—
Blue	Red	Green	6.2	—	—
Blue	Gray	Green	6.8	6.8	6.8
Violet	Green	Green	7.5	—	—
Gray	Red	Green	8.2	8.2	—
White	Brown	Green	9.1	—	—
Brown	Black	Blue	10	10	10
Brown	Brown	Blue	11	—	—
Brown	Red	Blue	12	12	—
Brown	Orange	Blue	13	—	—
Brown	Green	Blue	15	15	15
Brown	Blue	Blue	16	—	—
Brown	Gray	Blue	18	18	—
Red	Black	Blue	20	—	—
Red	Red	Blue	22	22	22
Red	Yellow	Blue	24	—	—
Red	Violet	Blue	27	27	—
Orange	Black	Blue	30	—	—
Orange	Orange	Blue	33	33	33
Orange	Blue	Blue	36	—	—
Orange	White	Blue	39	39	—
Yellow	Orange	Blue	43	—	—
Yellow	Violet	Blue	47	47	47
Green	Brown	Blue	51	—	—
Green	Blue	Blue	56	56	—
Blue	Red	Blue	62	—	—
Blue	Gray	Blue	68	68	68
Violet	Green	Blue	75	—	—
Gray	Red	Blue	82	82	—
White	Brown	Blue	91	—	—
Brown	Black	Violet	100	100	100

Table A-1   Nominal Resistor Values (Courtesy of the Allen-Bradley Company)

Appendix B

Equipment and Components

Standard Equipment for Basic Labs

1—Dual channel oscilloscope
2—DMM (Occasionally more are required)
1—Power supply, variable, regulated
1—Signal generator. (A function generator is required for a few labs.)
1—*LRC* meter or impedance bridge

Specialized Equipment for the Power Labs

2—Single-phase wattmeters (approximately 300 W full scale)
Three-phase ac source, 120/208 V

Resistors

Most resistor tolerances are not critical. However, unless otherwise noted, we recommend 5% tolerance resistors or better.

Resistors (1/4 W)

4.7 Ω, 6.8 Ω, 10 Ω, 15 Ω, 47 Ω, 75 Ω, 82 Ω, 100 Ω, 150 Ω, 180 Ω, 220 Ω, 270 Ω, 330 Ω, 470 Ω, 510 Ω, 680 Ω, 820 Ω, 1 kΩ, 1.2 kΩ, 1.5 kΩ, 2 kΩ, 2.2 kΩ, 2.7 kΩ, 3.3 kΩ, 3.9 kΩ, 4.7 kΩ, 5.1 kΩ, 5.6 kΩ, 6.8 kΩ, 7.5 kΩ, 9.1kΩ, 10 kΩ, 12kΩ, 18kΩ, 20 kΩ, 39 kΩ, 47 kΩ, 51 kΩ, 75kΩ, 100 kΩ, 180 kΩ, 330 kΩ, 3.3 MΩ, 5.6 MΩ, 10 MΩ

Resistors (Other)

(1/8 W)	470 Ω
(1/2 W)	47 Ω, 82 Ω, 470 Ω, 560 Ω
(1 W)	100 Ω, 220 Ω, 470 Ω
(2 W)	47 Ω, 75 Ω, 82 Ω, 100 Ω, 120 Ω, 270 Ω, 470 Ω, 510 Ω, 1 kΩ
(10 W)	10 Ω, 51 Ω
(25 W)	10 Ω
(200 W)	100 Ω (One for Lab 12. If you do the three phase labs, you will need three.)
(200-W)	250 Ω (Three. Required for the three phase labs only.)

Capacitors

4.7 nF, 2200 pF, 3300 pF, 0.01 μF, 0.022μF, 0.047 μF, 0.1 μF, 0.22 μF, 0.33 μF, 0.47 μF, 1.0 μF

Capacitors (Other)

10 μF, 220 μF, 470 μF, electrolytic
30 μF, non-electrolytic, rated for 120 VAC operation. (One for Lab 12. If you do the three phase labs, you will need three.)

Inductors

1 mH. Powdered iron core inductor (Hammond #1534A or equivalent)
2.4 mH (two). Powdered iron core inductor (Hammond #1534C or equivalent. You need an inductor with low resistance to approximate the behavior of an ideal inductor.)
0.2 H (approximate value) inductor. Rated to handle 2 amps.
1.5 H iron-core inductor. (Approximate value. Value not critical.)

Miscellaneous

2N3904 *npn* transistor, 741 op-amp, NE 555 timer, LM556C controlled oscillator
Voltage regulators (7805, 7905 and LM317)
Zener diodes (1N4773A, 1N4734A, 1N4735A)
LEDs
Thermistor, 10 kΩ @ 25°C
Diode: 1N4004 or equivalent
Potentiometers: 5 kΩ, 10 kΩ, 20 kΩ
455 kHz IF transformer
Transformers: 120/12.6 V (filament), 120/24 V, 120/30 (center tapped), Class2 transformer 12VAC, 500 mA (SPC9708 or equivalent)